RH
10
19
4/885
T

TECHNOPHOBIA

Also by Hal Hellman

Navigation—Land, Sea and Sky (1966)
The Art and Science of Color (1967)
Controlled Guidance Systems (1967)
Light and Electricity in the Atmosphere (1968)
The Right Size (1968)
High Energy Physics (1968)
Defense Mechanisms, from Virus to Man (1969)
The City in the World of the Future (1970)
Helicopters and Other VTOL's (1970)
Energy and Inertia (1970)
Biology in the World of the Future (1971)
The Lever and the Pulley (1971)
*The Kinds of Mankind: An Introduction to Race
    and Racism, with Morton Klass (1971)*
Feeding the World of the Future (1972)
Population (1972)
Energy in the World of the Future (1973)
*Transportation in the World of the Future (1968;
    second edition 1974)*
*Communications in the World of the Future (1969;
    second edition 1975)*

Hal Hellman

TECHNOPHOBIA

Getting Out of the Technology Trap

M. Evans and Company, Inc. / New York, N.Y. 10017

M. Evans and Company titles are distributed in
the United States by the J. B. Lippincott Company,
East Washington Square, Philadelphia, Pa. 19105;
and in Canada by McClelland & Stewart Ltd.,
25 Hollinger Road, Toronto M4B 3G2, Ontario

LIBRARY OF CONGRESS CATALOGING IN PUBLICATION DATA

Hellman, Harold, 1927-
    Technophobia: getting out of the technology trap.
    Bibliography: p.
    Includes index.
    1. Technology. 2. Technology assessment. I. Title.
T49.5.H44   301.24′3   75-44372
ISBN 0-87131-206-9

Design by Joel Schick

Manufactured in the United States of America

9 8 7 6 5 4 3 2 1

For Sheila, my inspiration

CONTENTS

ACKNOWLEDGMENTS

Research for a book like this is done mainly in the world of printed matter. Some four thousand index cards, two file drawers stuffed with notes, tear sheets and photocopies, and several bookshelves filled with books, magazines, reports, and pamphlets will attest to that. To those who generated this information, and to those—librarians, public information people and others—who helped me amass this material, I am deeply indebted.

An important additional resource, however, consisted of people, both colleagues and strangers, who were willing to share their knowledge with me by way of personal interviews. Among them I would like to single out:

Dr. Marvin L. Aronson, director of the Group Therapy Department at the Postgraduate Center for Mental Health, practicing psychotherapist, and author of *How to Overcome Your Fear of Flying;* Dr. René Dubos, eminent microbiologist, experimental pathologist, and author of the Pulitzer Prize winner *So Human an Animal* as well as several other fine books; Mr. Charlie Hance, potter who turned Brooklyn in for Maine; Mr. Earl Horn, a gentle college student from New England; Ms. Sheila Klass, author and Professor of English at Manhattan Community College; Dr. Arthur O. Lewis, Jr., Associate Dean at Penn State, ex-chairman of the Science, Technology and Society Program, and editor of *Of Man and Machines;* Mr. Charles Mann, curator of the rare book room at Penn

State. Dr. Hugo Meier, Professor of History at Penn State; Fran Merritt, creator and director of the Haystack Summer Crafts school on Deer Isle, Maine; Dr. Philip Morrison, Professor of Physics at M.I.T. and book review editor for *Scientific American;* Dr. Don K. Price, dean of the Kennedy School of Government at Harvard University and author of *The Scientific Estate;* Mr. Wes Thomas, computer specialist and editor of the futurist publication *Synergy/ Access;* Dr. Daniel Walden, Associate Professor of American Studies at Penn State; Dr. Joseph Weizenbaum, Professor of Computer Science at M.I.T.; and Dr. Jerome Wiesner, President of M.I.T.

I wish to express my appreciation to my wife Sheila, who not only suffered with me through the long gestation time of the book, but sat through or participated in innumerable discussions and arguments regarding technology and its place in our society.

A special set of thanks goes to Dr. Jerold Schwartz, Assistant Professor of Psychology at Montclair State College, and Dr. Norman Dane, Professor of Intellectual History at Rutgers University, each of whom read parts of the manuscript and helped clarify certain unclear points.

My final, and main, tribute goes to Mr. Leonard L. Lederman, director of the National Science Foundation Office of National R & D Assessment, and Dr. Morton Klass, Professor of Anthropology at Barnard College, Columbia University, both of whom not only read through the entire manuscript and offered extensive comments and criticisms, but also were willing, time and again, to discuss these or additional points.

I am also indebted to my students at Fordham and Fairleigh Dickinson Universities, where I taught courses based on the ideas contained herein, who also provided grist for my mill.

PREFACE

Modern society is in deep trouble—what with inflation, recession, shortages, overpopulation, crime, and pollution —and many people think the blame lies with our way of life: they think we have created an industrial, mechanistic, technological monster.

So maybe, as is often suggested, the answer is to go back to a state of nature. Should we simply repudiate technology? Do away with it? Should we stop it where it is? Go back to 1900? 1700? 10,000 B.C.? Is technophobia—by which I mean fear of technology, fear of science, fear of change in general—justified?

We didn't always think so. Belief in technological progress was widespread in the nineteenth century, and the twentieth brought the real and important successes of our scientific and technological effort during World War II. And then, in 1957, Sputnik stimulated great excitement and interest.

The marvels of technology were all too evident—radar, computers, transistors, television, and space flight! Surely they could do everything: clean up our cities and give everyone a job; make everyone healthy; even feed, clothe, house, and educate everybody. Utopia was just around the corner.

But we didn't reach Utopia, the reasoning, only slightly simplified, going as follows. Technology has increased and so have our problems, ergo, technology is the cause of our problems. For reinforcement we had the Vietnam War. Napalm is bad; napalm is technology; therefore technology is bad. Some of the current anger about technology has to do with an incomplete understanding of what it is and its significance in modern society. The term *technology* is often taken to be synonymous with *machines*, but this is much too limited a usage. Some of the onus is removed if technology is thought of as the application of scientific knowledge to the meeting of human desires and needs.

But I would say that much of the anger also derives from a quite different source. Our large technological capability forces us almost to the brink of realizing that the causes of most of our problems are social. We are the enemy. And who can face that?

This is not to say that there is really nothing to fear from technology, but rather that there are rational fears and there are irrational fears.

Watch a copying machine in a busy place for a while. Although it is no longer a new invention, it is still large enough and new enough to scare people. Many using it for the first time approach it warily, as if it is some kind of untrustworthy beast that will, if given half a chance, leap out and bite them. That the machine has been made as simple and safe to operate as possible, that complete step-by-step operating instructions are prominently displayed, seems to make no difference.

For three full years a friend of mine battled with his car, a basic import with manual choke and transmission. I drove it a couple of times and found it to be a fine little automobile.

On several occasions I drove with him and discovered some interesting things. When starting the car he gave it so much gas that he flooded the engine almost immediately; somehow he always had the choke too far in or too far out; when driving he was almost never in the right gear; and while stopped for a light, he let his left foot rest comfortably on the clutch.

I suggested, as delicately as I could, that perhaps some of the car's trouble lay with him. I even tried giving him some pointers on use of the choke one winter morning, taking into account the various weather and driving conditions, but he quickly lost patience. "I can't remember all that stuff," he fumed.

I tried to get him to listen to the engine. But there was something in him that just wouldn't permit this either. It would be giving in to a machine, and he simply refused. So he went back to cursing the car, automobile manufacturers, the service station, and technology in general.

One way out for my friend was to move back to the city from the suburb in which he lived, and give up the car altogether. But this would have meant giving up a beautiful studio in his house. Another alternative was to buy a car with an automatic choke and automatic transmission. He ended up with a more expensive car, which he could not really afford, and which had the additional drawbacks of being more complicated (more things to go wrong), and giving lower gas mileage!

It would obviously have been better if he had been able to come to terms with his first car.

Like it or not, this is a technological world, and it requires at least some technological literacy. Our homes are filled with electrical appliances. Permitting people who have no basic understanding of electricity to plug into the electrical system is wrong and dangerous, and may explain a large part of the thousand or so electrocutions and the many more electrically caused fires that take place each year.

A professor of political science I know was teaching a

course on technology and society, yet he couldn't assemble a set of electric trains for his youngster because he didn't know that electricity must move in a complete circuit. I call such a person a *technological illiterate*.

Even without the basic "personality conflict" involved between technology and technophobes, assimilation of technological illiterates into modern society becomes virtually impossible because of the extraordinary idea still extant that a person can be called intelligent who knows nothing of science and technology.[1] Indeed, we may be living in a scientific age, but a recent study shows that our younger generation's knowledge of science is going down! [2]

Technological illiteracy can just as easily lead to a blind, uncomprehending reverence for science and technology as to an anti-intellectual fear of them. Both—and we have had too much of both—are equally bad.

The choice, then, is not whether to take an ax to the next machine you see or to tie a red ribbon around it. Rather it is a matter of deciding whether—and if so, how —to live with it.

For those of you who are not sure where you stand, I have created a short questionnaire that should give you a better idea of your own feelings on the matter. For the first four questions, score +1 for each "yes," −1 for each "no," and 0 for "don't know."

Are You a Technophobe?

1. Does the front door lock seem to freeze up as soon as you insert your key, while it remains warm and inviting, and quite operative, to everyone else in your family?

2. Do drawers, doors, and windows seem to stick only when you attempt to open them? (If they stick for everyone, score 0.)

3. Do watches, clocks, irons, and other small mechanical or electrical devices seem to break as soon as you put them on, pick them up, or whatever?

4. Do you approach a new machine, such as a new car or a copying machine at the library, with fear and trepidation?

5. Do you steadfastly refuse to read the step-by-step instructions on such machines, or on new appliances, before operating them? (If you get some one else to do the work for you, or to direct your efforts, score +1.)

6. How do you feel when you see a large industrial complex, or a wheat field being harvested by a row of machines? (For panic or annoyance, score +1; pride or interest, −1; indifference, 0.)

7. Dr. Ruth M. Davis, director of the Institute for Computer Sciences at the National Bureau of Standards, has said, "Computers will make possible the mechanization of intelligent behavior to an extent that is essentially limitless." What's your reaction? (Fear or alarm, +1; excitement/challenge, −1; indifference or no opinion, 0.)

8. How do you feel when you see a picture of a nuclear power plant, or read about the next step, breeder reactors? (Pride and interest, unmixed with fear or worry, score −1; total panic, +1; indifference, don't know, or neither, 0.)

9. Any problem, social, psychological, or economic, can be solved by science and technology, or at least by being perfectly rational about it. (True, −1; false +1; don't know, 0.)

All right. Now add up your score. If you can't manage it, add another +1 and give it to some one else to do.

If your score lies between −3 and −7, you are a technophile, but with a heart and brain.

If your score is −8, −9, or −10, you are a technological rhapsodist. That means you are dangerous to society and should be locked up or reprogrammed.

If your score lies between −2 and +2, you are also a problem, though less than the rhapsodist, for it suggests a lack of interest in a subject that has a major bearing on your life. The fact that you are reading this book is hopeful, however.

If your score is +3 to +5, you are a second-class technophobe; no major therapy is indicated.

If your score ranges anywhere from +6 to +11, you are a first-class technophobe. There are five ways out for you:

1. Continue on as usual, namely, complaining about how technology has ruined your life and the lives of those around you, regardless of whether *they* feel they are living ruined lives.

2. Get someone who is "handy" to go through all your troublesome motions with you. Wherever necessary, (s)he will put graphite in your locks and silicone on your window slides; plane swollen wood, etc.

3. Go to engineering or at least trade school, though this is admittedly a rather extreme measure.

4. Try psychotherapy. Maybe you were frightened by a toy train when you were very young.

5. Read this book twice.

1 *Future Shock*

A brash young Texan, come to New York to do some sightseeing, hired a taxi for the day at the airport. On the way into the city, the cabby proudly pointed out Shea Stadium.

"Not bad," said the Texan. "How long did it take to build?"

"Oh, couple of years, I guess."

The Texan snorted. "We've got a bigger one and we got it done in less than a year."

They were now approaching the Triboro Bridge. "Hey," the Texan commented, "that's a nice bridge. How long did it take to build it?"

The cabby was wary by now. "Six months."

"Shucks. We built one just like it in three months."

A little later, as they were driving up the West Side Highway, they passed under the twin 110-story towers of the World Trade Center. "Wow," exclaimed the Texan. "How long did it take to build that?"

"Can't say for sure. But it wasn't there this morning."

In 1965 the term *future shock* began to enter the public consciousness. It was coined by Alvin Toffler and used in an article in *Horizon* Magazine to describe "the shattering stress and disorientation that we induce in individuals by subjecting them to too much change in too short a time." Five years later his book with that title seemed to touch the nerve center of the industrialized world, and that of the aspiring countries as well. The book has been reprinted many times and translated into a number of languages.

Everything around us, wrote Toffler, keeps changing so rapidly we can't keep up with it. We are living in the midst of a wildly speeded-up kaleidoscope that is making us dizzy, anxious, or worse.

Toffler wrote of the transitory nature of human ties in urban society; the fact that we can rent almost everything, including people; he showed us the throwaway society, including the rapid construction and tearing down of buildings!

Rate of Change

"In the three short decades between now and the twenty-first century," wrote Toffler, "millions of ordinary, psychologically normal people will face an abrupt collision with the future." He suggested that the rate of change is accelerating at a pace that far outstrips anything ever experienced before.

But let's look at that word, *acceleration*. Say that our society is experiencing ten changes a week—however we define change. At the end of two weeks we will have experienced twenty changes, at the end of three weeks, thirty, and so on. We are indeed experiencing increasing change, but the *rate* of change—ten changes a week—remains constant.

But if we are experiencing ten changes one week, eleven the next, twelve the next and so on, then the rate of change *is* accelerating (increasing) at the rate of one change per week, and after three weeks we will have experienced thirty-three changes, not thirty. And we will continue to whirl faster and faster until, at some point, we explode from pure centrifugal force.

This, say the future shockers, is where we are, or seem to be heading. If it's where we are, however, one wonders why we haven't all been locked up already. If it is where we are heading, at least there's comfort in the fact that trend is not certitude: there is nothing that says the rate has to continue to increase. I feel there is even some reason to suspect it will decrease.

At the very least, the percentage of people working in the fields of science and technology seems no longer to be increasing faster than total population, as it has in the past couple of decades. A few years back someone calculated that if the rate had continued to increase at the same speed, then by some time in the next century everyone in the country would be a scientist—which doesn't seem very likely. The rate of increase has slowed markedly in the last couple of years—and this is not because we have reached some kind of saturation point with respect to scientific and technical personnel. Dr. D. S. L. Cardwell, a specialist in the history of science and technology, tells us that "during the years 1790–1825 France had more scientists and technologists of the first rank than any other nation has ever had over a comparable period of time; on a per capita basis the French record has not been equalled even up to the present day." [1]

And there is an even more basic question: Is there a constantly growing shoreline of knowledge? There is by no means full agreement that the answer is yes. Gunther S. Stent, professor of molecular biology at the University of California, argues not only that scientific progress is a self-limiting process, but that it is rapidly nearing its end. He points out, for example, that progress clearly has *not* been a built-in factor operative throughout man's history.[2]

Even the question of just how fast change is actually taking place today is not as easy to answer as it has been made out to be. What we have really is a *perception* of change that seems to be overwhelming. But of what does this change consist? How, for example, does one handle change that is periodic, or pendular? If a fad is introduced but then, due to the innate good sense of most people, falls by the wayside, is that double change, or zero change? Consider one of Toffler's examples, that of paper clothing. He reported: "Fashionable boutiques and working class clothing stores have sprouted whole departments devoted to gaily colored and imaginatively designed paper apparel. Fashion magazines display breathtakingly sumptuous gowns, coats, pajamas, even wedding dresses made of paper."[3]

Yet today, five years later, one hears little of this. It was a fad. Toffler suggests that fads are evidence of (a cause of) future shock. Perhaps that is based on the idea that each fad has a double effect—once coming in, once going out. But actual change that results is zero, and that is what counts.

Toffler wrote: "We are moving swiftly into the era of the temporary product, made by temporary methods, to serve temporary needs."[4]

But (and Toffler may smile at this) times have changed. We may be moving *out* of the era of the temporary product, and back to the days when cars, refrigerators, and clothing were not discarded just because newer models had come in. Is this a point in favor of future shock? Two points? Or none? Irving Louis Horowitz, professor of sociology and political science at Rutgers University and editor-in-chief of Transaction/Society, makes a similar point: "Labels have changed with breathtaking suddenness, but systems have changed much more slowly." [5] He points out, for example, that "despite the convulsions of the last decade, progress can be very slow. At the beginning of this century the United States, Russia, Japan, Western Europe and England—the big five of Imperialism—dominated the world. Now after two world wars, countless numbers of mini-wars, endless social confrontations, international realignments and permanent cold wars, the list is the same." [6]

Now vs. Then

The perception of rapid change is particularly acute when the real today is compared with an imaginary yesterday. The common idea of earlier times is almost invariably that of bucolic sleepiness. In earlier days, we often hear, the life led by one generation was almost exactly the same as that of the next.

Indeed, the whole point of future shock is that change is occurring at an unprecedentedly high rate, which surely implies that things were much quieter before. Toffler quotes C. P. Snow to back up his feeling that before this century social change was "so slow, that it would pass unnoticed in one person's lifetime. That is no longer so. The rate of change has increased so much that our imagination can't keep up." [7]

But as J. D. Bernal, British historian of science and technology, has suggested, "In analyzing the past there is always the danger of projecting present ideas backward." [8] The nineteenth century, for example, seems to many of us a time of peace and constancy, particularly with respect to our own, a time when science and technology had not yet reared their ugly heads, a time to which many of our present-day young would like to return.

Yet in many ways the nineteenth century was quite as explosive as our own, and perhaps more so. Consider transportation. A practical river steamboat appeared in 1807. In the half century after the War of 1812 the country went canal crazy, constructing 4,000 miles of these artificial waterways in the eastern United States alone. By 1825 the railroad made its appearance here; between 1830 and 1860 more than 30,000 miles of railroad track were put down! To give you some idea of what that meant, consider that our own interstate highway system, which will when completed consist of 42,500 miles of roadway, has been called the greatest public works project the world has ever seen.

Indeed, the railroad provided a pacing factor for industrialization. But even more shattering, in terms of change, is what Leo Marx refers to as "the machine in the garden." In his excellent book of that name he says that "nothing quite like the event announced by the train in the woods had occurred before. . . . For in the stock contrast between city and country each had been assumed to occupy a more or less fixed location in space: the country here, the city there. But in 1844 the sound of a train in the Concord woods [where Thoreau stayed and wrote his book, *Walden Pond*] implies a radical change in the conventional pattern." [9]

It was in the nineteenth century that land travel broke out of bonds that had held tight since horses were first used to carry men. The same maximum speed—perhaps twenty miles an hour over shorter distances and down to about eight miles an hour for long-distance relay systems like the old Pony Express—had held steady for some two thousand years. The development of the train was surely a break equivalent to, if not surpassing, our breaking of the sound barrier. By the end of the century, trains had reached speeds of a hundred miles an hour. Except on a few short runs, not a single American train exceeds this speed today, though some foreign ones do.

Steam-propelled ocean liners and airships, bicycles, automobiles, and experiments with heavier-than-air craft round out the century's developments in transportation.

Invention of the elevator and steel-frame construction combined to make the skyscraper a reality. Ada Louise Huxtable, architecture editor of the *New York Times*, maintains that "The twentieth century owes the 19th century more than it can ever acknowledge. And it is both thoughtless and churlish about repaying the debt." [10]

Change? The nineteenth century saw the dethronement of land. By the end of the century, the country's leaders were no longer the landed aristocracy; the men of enterprise had taken over. Big business, in other words, is by no means a twentieth-century invention. Nor, even, are the big research labs we hear so much about. These started in Germany around the 1860s, inspired to a great extent by Justus von Liebig's chemistry laboratory. Dr. Bernal tells us that the nineteenth century "was predominantly the century of transition for science. It was to change during those hundred years from an elegant ornament of society, practiced by virtuosi, to an essential factor in the everyday production of goods and services." [11] Included herein would be America's agricultural colleges and research stations, which put agriculture on a scientific basis and made it possible for the United States to become the "breadbasket for the world."

Alfred North Whitehead, the eminent British mathematician and philosopher, summed it all up this way: "The greatest invention of the nineteenth century was the invention of the method of invention." [12] The result was an utter torrent of new devices and new ways of doing things.

Synthetics, for example, are a nineteenth-century development. Artificial silk, the forerunner of nylon and rayon, dates back to 1884; cellophane and celluloid (the first "plastic"), to the 1860s; analine dye, to 1856; and urea, to 1828. The synthesis of urea was a truly significant development, for this was the first time that an "organic" substance (urea is found in urine) was created by man. It was the first indication that living things were not made of some special kind of matter, a major factor in understanding the workings of the body.

The development of interchangeable parts in manufacturing can be dated back to the fifteenth century and the printing press; the implementation, as in the manufacture of rifles, was definitely nineteenth century, however, and an important aspect of industrialization.

Change? Population in the United States leaped from 12 million in 1830 to 75 million at the turn of the century, a six-times increase. Then, over the next seventy years the population grew to 200 million, an increase of some two and a half times. By 1890 the frontier had been pushed right into the Pacific Ocean.

Prior to 1800, whatever industry there was, was in shops where ten or twelve craftsmen of various skills worked by hand. The vast majority of people worked on the land. In 1839, 72 percent of the labor force still worked in agriculture, only 17 percent in manufacturing. By the end of the century the figures had reversed, to 33 percent in agriculture and 53 percent in manufacturing.

Fed by labor released from the land, and abetted by industrial concentration and improved land transportation, cities grew rapidly, even chaotically. Between 1800 and 1900 New York grew from 79,000 inhabitants to 3,437,000;

London, from 959,000 to 4,536,000; Paris, from 548,000 to 2,661,000; and Berlin from 173,000 to 2,712,000!

There you have change!

Literacy also grew rapidly. Early in the century reading was a skill possessed by a minority of Americans; by 1900 almost 90 percent of the population ten years of age and over was literate.

With the introduction of the typewriter, telephone, and other kinds of office equipment came much new employment opportunity for women. The fight for women's suffrage started just about mid-century, and emancipation was a nineteenth-century accomplishment.

Some other nineteenth-century developments? The internal combustion engine and diesel engine, practical turbine engines, machine tools, fully automatic manufacturing processes, ball bearings, division of labor. Electricity (plus electric bells, ironers, cookers, lights, motors, etc.), refrigeration, the phonograph, radio, moving pictures, subway and street cars, the transatlantic cable, telephone and telegraph, Morse code. Mechanical harvesting equipment, chemical fertilizers, canning, automatic clothes- and dishwashers, vulcanization of rubber, pneumatic (air-filled) tires, ship propellers. Photography, organic chemistry, penny post (mail), pneumatic post, the first cheap newspaper in English, an international news agency, the first oil well, steam-rollers, the spectroscope, pasteurization, antiseptic surgery, anesthesia, aspirin. Carpet sweepers, synthesis of fats, central heating, automatic calculators, cash registers, Linotype, photoelectric cells, and the friction match.

Add to these the harmonica, saxophone, zipper, wood pencil with rubber eraser, and safety pin and one wonders how the nineteenth-century American survived!

And even the number of changes do not tell the whole

story, for the early stages of change are probably more traumatic and dramatic than the later ones. Consider, for example, the furor about the SST or supersonic transport, which was going to revolutionize air travel. A standard commercial jet takes about six hours to cross the Atlantic Ocean. With the Concorde, the British/French entry into the field, it takes about three hours. And with the American design, which is not being built, it would have taken two hours.

Thus we can say that the American SST would have flown three times as fast as a modern subsonic jet, which sounds like an extraordinary, shocking change. But the *savings*—for the elite few who would have used it—would have been four hours. Now put yourself back in the nineteenth century. Crossing the country by horse and wagon took an average of six months; by ship around South America it took three months. But in 1869, when the last spike was driven to complete the first transcontinental rail line, the time was cut to six days. The point, of course, is that not only was the railroad thirty times faster than the previous land method, but the saving was more than five and a half months!

How many twentieth-century developments is that worth?

Or consider communications. We hear continually about the communications revolution. But radio is a nineteenth-century development, as are the phonograph and even the movies. Television admittedly has added an important dimension to radio (though the first experiments date back to the last century). We have seen color television in the last few years. We may see two-way, interactive television, and maybe even three-dimensional television in the next decade or two. But I would call all these relatively minor changes.

I am not trying to downgrade the admittedly powerful effects of television. It, along with the many other twentieth-century developments, has surely altered our lives in many ways. Television does provide a stronger impact than radio because of the visual information it supplies. But it is important to compare it to what there was before it came into being; the telegraph, telephone, radio, and talking pictures had all been around for years.

The real impact of the broadcast media is immediacy; it can tell you what is happening right now. But from the standpoint of immediacy in transferring basic information, the telegraph accomplished that almost as well—and particularly so when compared to what there was before. Prior to the introduction of the telegraph in the mid-1800s, it might have taken months before a farmer heard who the new president was, or who won some championship bareknuckle fight, or that we were at war. Indeed, the last battle of the War of 1812, the defense of New Orleans at the cost of a thousand men, occurred two weeks after the peace treaty had been signed in Belgium! Unfortunately, the only means of communication between Europe and the United States was by ship, and the trip across took a full month.

Before there were telegraph stations and telephones to communicate interest in the affairs of the country, the farmer, in a very real sense, was a citizen of the United States in name only. And I would venture to suggest that the major scientific theories of the nineteenth century actually had a greater impact on the average citizen than have those of the twentieth. In the twentieth century we have had the relativity theory, the uncertainty principle, the nuclear atom, quantum mechanics, and, perhaps, elementary particle theory. It seems to me that these are easily matched, in terms of the impact on everyday life,

by the basic principles of thermodynamics, of electro-
magnetic field theory, the periodic table of the elements,
the germ theory of disease, the medical model of mental
illness, and Darwinian evolution. Again, I'm referring to the
impact on everyday life and technology.

You may counter that that's because the earlier theories
were more basic or simpler or more fundamental. I answer
that that's exactly the point.

And, with all that, we are still coping.

Breakthroughs

I once asked an editor of *Scientific American* why their
articles were often so difficult to read. His answer, only
half in jest, was that all the simple subjects had been
used up. I wonder if that doesn't reflect the actual situation
in scientific and technological research. We have entered
a more complex era in which each worker claims a revolution, a breakthrough, while too few science writers and re-
porters know enough to realize that each breakthrough
has probably taken place several times before.

Now, I do not wish to underplay the truly awesome
possibilities opened up by, let's say, developments in
molecular genetics—control of the sex of newborn infants,
control of genetic disease, possible changing of our genetic
heritage, cloning (making multiple copies of a single organism), and so on. But with each tiny development in cloning the same apocalyptic stories—10,000 Hitlers or even,
smugly, Einsteins—are dredged up all over again. Yet
scientists are still in the carrot and frog stage.

Even if a real step is made, the road from there to
actual practice may be long and arduous. I must have read
five times in the last five years that Dr. So-and-so's work
suggests we will be able to use the protein locked up in

green leaves. This would have tremendous significance for the tropics, where many of the nutrients needed by humans are found not in the soil but in inedible vegetation. But the new methods to extract leaf protein haven't come anywhere near being put into practice. It's all totally uneconomical and still highly experimental. The galloping pace is a mirage, a gross exaggeration.

We are also filled with "new" ideas in public transportation, such as magnetically suspended trains, moving sidewalks, and so on, a few of which have actually been built. Yet a Professor A. C. Albertson proposed a magnetically suspended train back in 1903; there was even a drawing of it in *Harper's Weekly*. There was also an illustrated proposal for underground moving sidewalks in New York. The caption reads:

> An interesting phase of New York's transportation problem is the proposal to construct underground moving sidewalks or platforms. The first one planned, of which the details are shown in the above drawing, is to go from Williamsburg to Bowling Green, and is to connect on the way with the surface and elevated cars. The fare will be one cent, and the speed will be from five to nine miles an hour.[13]

Note the optimistic use of the terms "is to go" and "is to connect." It's now three-quarters of a century later and we're still in the talking stage.

In his book, *Automobiles of the Future*, Irwin Stambler wrote in 1966: "The trouble with today's world is that tomorrow arrives too soon. At this writing, there are less than 100 gas turbine cars on America's roads. Within a few years, these numbers will go up to thousands and, by the 1970s, probably millions."[14]

Now that we have finally begun to worry about the supply of gasoline for our precious cars, we hear suggestions that alcohol be used, and derived from vegetation. But the Germans did that on a large scale in World War II, and even fooled around with the idea at the turn of the century.

I wonder, then, whether the inundation we feel is not one of information, rather than real change. "Several decades ago," writes L. Stevenson in *Science News*,

> we saw brilliant ideas in quantum theory and relativity confirmed by ingenious experiments. Nuclear energy and transistors have also opened vast new areas in technology. In recent years, however, big research budgets have produced many big machines but not many big results. In contrast to the earlier successes in the theoretical predictions and experimental confirmations of photons, positrons, mesons, etc., we have seen futile searches for quarks, monopoles, gravitons, tachyons etc.[15]

Perhaps what we have been experiencing is not future shock but information, or sensory, overload.

Information Overload

And even here things may be somewhat overstated. The current idea, again, is that we simply cannot keep up with the increasing amount of information being dumped on us. Prof. Philip Morrison once suggested a fascinating image: If the accumulation of information is thought of in terms of books on a shelf, then we are reaching the point where the end of the shelf is moving away from us faster and faster, and may soon reach the speed of light!

But what, really, is all that new information? And how much of it is important?

For we do not only accumulate information and knowledge on a bit-by-bit basis. Usually what is needed is some kind of unifying theme.

An interesting analogy is seen in the chess played by grand masters as opposed to that of the rest of us duffers. Experiments show that the average player, given a chance to study a position on a board for five seconds, can reconstruct positions containing no more than five or six pieces; the grand master, by contrast, can reconstruct positions with as many as twenty-five pieces!

The reason has less to do with good memory than with the fact that the grand master has studied positions so much that he tends to see patterns rather than individual pieces. When the pieces were placed at random on the board—so that they did not represent the typical positions of a game—the grand masters' results fell right down to the level of the amateur!

Dr. Herbert A. Simon, professor of computer sciences and psychology at Carnegie-Mellon University, suggests that the number of patterns mastered by the grand master is roughly the same as the number of words mastered by the average intelligent person, or about fifty thousand.

The vast multitude of facts and information with which we are deluged creates a difficult but not impossible burden, for we now have more theories (patterns) to help us. The periodic table of the elements eliminates the need to "know" all of the elements—that is, to keep in mind a mass of irrelevant facts—while quantum mechanics has provided a theoretical basis for that information. Anything we want to know can then easily be figured out or, of course, looked up.

Those in the high-energy physics (elementary particle) field, or cancer researchers, surely do not feel that they are inundated with information. They are eagerly seeking more.

So we are not so hopeless, helpless, or hapless as some would have us believe. And if we cannot be Renaissance men and women, let us appreciate the fact that anyone who wants to can be smarter in some ways than the most brilliant scholars of yesteryear—for we can learn what they knew without being burdened with much of the incorrect information they also had.

Innovation and Diffusion

Daniel Bell, well-known sociologist and futurist, suggests that it may not be the rate of technological invention that has picked up so much as the rate of diffusion through the economy. He mentions a study done by Frank Lynn for the President's Commission on Automation that showed that the average time between initial discovery and recognition of commercial potential has been decreasing:

1880–1919	30 years
Post-WWI	16 years
Post-WWII	9 years

Also, he says, the time required to translate a basic technological discovery into a commercial product or process decreased from seven to five years during the sixty- to seventy-year time period investigated.[16]

The idea is widely held. But there is no universal accord on it. Edwin Mansfield, who did an extensive study of the diffusion of innovation in industry, maintains that even the rate of adoption does not appear to have increased as time goes on.[17] Support for his position is given in the table shown here, which shows the time lag between conception of a new idea and its successful innovation as a usable product or process. The study, done by the Battelle Columbus Laboratory for the National Science Foundation, shows that there is no pattern of shortening as time goes on.[18]

TABLE 1

DURATION OF THE INNOVATIVE
PROCESS FOR TEN INNOVATIONS

Innovation	Year of First Conception	Year of First Realization	Duration (years)
Heart pacemaker	1928	1960	32
Hybrid corn	1908	1933	25
Hybrid small grains	1937	1956	19
Green Revolution wheat	1950	1966	16
Electrophotography	1937	1959	22
Input-output economic analysis	1936	1964	28
Organophosphorus insecticides	1934	1947	13
Oral contraceptive	1951	1960	9
Magnetic ferrites	1933	1955	22
Video tape recorder	1950	1956	6
Average Duration			19.2

One reason for the disparity in results has to do with how we define the point of invention. In a study made several decades ago, the sociologist S. Colum Gilfillan gave this example:

Suppose we consider television. That sounds simple, since we all know what television is. Or do we? Was it already television in 1847 when Souvestre satirically predicted it? Or did it begin in 1877 when the first apparatus was built, or in 1882 when the scanning disk was added, or in 1901 when Fessenden designed a wireless system? Or was it Zworkyin's modern cathode ray receiver of 1929 that constituted the invention of television, with the Kinescope, and some experimental broadcasts the next year? Or is our date 1928 or 1937 when regular broadcasting began? [19]

Depending on which date is chosen, the rate of change can be shown to be shorter or longer.

Again, all of this is not to say that rapid change is not taking place. It clearly is. But it is also true that the advances so feared and spoken about by the popular press are usually far in advance of our actual capability. What is conceived as possible one day is reported as imminent the next, and accomplished on the third. In reality the basic changes—automobiles, jet aircraft, television, data banks, and so on—are introduced relatively slowly.

This being so, we find that our social system really has more time to keep pace than it appears to have.

Through the Ages with Future Shock

The perception of a speed of change that is outpacing the capability of society to keep up with it is hardly a new one. It is an idea that keeps cropping up in history. René Dubos says that "future shock occurs at *any* time in the course of history when there are some fairly rapid social or technological changes, and that has happened several times." [20]

He gives as an example a small book published just four hundred years ago by Louis Le Roy, the English title of which is *Of the Interchangeable Course and Variety of Things in the Whole World.* Basically, Le Roy was worried about the effects that the physical and intellectual developments of his time were having and would have on society. As has happened with Toffler's book, it seems to have touched a nerve center, for it was reprinted six times in the next decade and translated into both Italian and English.

Like ours, his was an era of great explorations. It saw the opening up of the ocean highways, discoveries of the vastness of our earth, conquests all over the world. Europe's interests were turning outward. Relaxation of sexual codes plus contacts with people far and wide led to rapid spread of new diseases, including syphilis and other venereal diseases.

It was a time when both printing and firearms came into widespread use. Breakdowns in ages-old codes of belief and action—in state, church, and home—resulted from the rapidly increasing use of the printing press. One can hear clearly the laments, "What is the world coming to?!" The widespread introduction of gunpowder and firearms made all the standard weapons obsolete. If a man's castle is no longer his fortress, what's left?

"All is pell-mell, confounded," Le Roy wrote, "nothing goes as it should." [21]

Nineteenth Century Perceptions

That was sixteenth-century future shock. With respect to the nineteenth century, we have already looked at what was actually happening; let us look now at some of the perceptions of what was going on at the time. This is quite as valid a procedure as actually counting changes. In voodoo, for example, it is surely the perception (that is, the fear) of death or injury that makes the magic work. In the same sense, if there is a perception or feeling of future shock, then it exists. And for just about every objection we hear today concerning science, technology, industrialization, urbanization, and progress, I was able to find, in the rare book room at Penn State University and elsewhere, one or more comments to the same effect. For example:

Loss of intimacy: "You may regret that you can no longer find a second home in the 'snug' little hotel, where the landlord was an old friend of the family and the waiter called you Master Harry to the last. But the times are against you." [22]

Too big: "We do things on a much larger scale than our forefathers. We have bigger ships, bigger houses, bigger shops, bigger dinners, bigger everything." [23]

Submergence of the individual: "The individual withers, and the world is more and more." [24]

Mastery of nature: "Our true deity is Mechanism. It has subdued external Nature for us, and we think it will do all other things. We are Giants in physical power; in a deeper than metaphorical sense we are Titans, that strive, by heaping mountain on mountain, to conquer Heaven also." [25]

Our time is unique: Senator Daniel Webster, in 1847, said of that time, "It is an extraordinary era in which we live. It is altogether new. The world has seen nothing like it before. . . ." [26]

Change: In 1872 Gustave Doré and Blanchard Jerrold, in the introduction to their book, *London. A Pilgrimage,* tell us that London's "passion for the New is shown in the hundred changes of every passing hour."

Some other comments:

1819: Washington Irving, in "The Legend of Sleepy Hollow," writes of "the great torrent of migration and improvement, which is making such incessant changes in other parts of this restless country . . ."

1829: Thomas Carlyle complains, in "Signs of the Times," that "the time is sick and out of joint . . . a boundless grinding collision of the New with the Old." [24]

Shock Treatment

The early years of the nineteenth century, particularly in the England of 1811–16, were difficult times. The hand-loom weavers were among the worst sufferers, because of the appearance of new spinning and weaving machines; nor were there any attempts to help them over this great change in their work and condition. There were many riots, some led by a man named Ned Ludd. (As a result, a person who strongly opposes progress, and particularly new machinery, is called a Luddite.) The riots were partly attempts to smash the machines that the weavers blamed for their troubles, and partly protests against food scarcity and unbelievably harsh working conditions.

The riots, though generally put down with great force, continued for several years, and did call the attention of reformers to the workers' plight. The riots tended to cease with the return of more prosperous times.

In 1830 a group of Thimonnier's wooden sewing machines were destroyed in a fire set by Paris tailors who were fearful of the new competition. They were right, of course. The sewing machine made it possible for almost anyone to sew well, including housewives, and was to revolutionize the manufacture of clothing and thereby the styles in all the civilized countries. Thimmonier, by the way, had to flee for his life.

The Machine in the Garden

Even the precursors to the modern automobile ran into trouble. Here is an account of a ride taken in 1829 in one of the steam-propelled coaches made by Sir Goldsworth Gurney.

We numbered four in the coach attached to the steam carriage. . . . Upon our arrival at Malksham, we found that there was a fair in progress, and the streets were full of people. Mr. Gurney made the carriages travel as slowly as possible, in order to injure no one. Unfortunately, in that town the lower classes are strongly opposed to the new method of transportation. Excited by the postilions,* who imagined that the adoption of Mr. Gurney's steam carriage would compromise their means of livelihood, the multitude that encumbered the streets arose against us, heaped us with insults, and attacked us with stones. The chief engineer and another man were seriously injured. Mr. Gurney feared we could not pursue our journey, as two of his best mechanics had need of surgical aid. He turned the carriage into the court of a brewer named Alex, and during the night it was guarded by constables.[28]

Although it is not generally known, steam carriages continued to be built and experimented with right through the century, sometimes with considerable technical success. But they frightened horses as well as people, and the times were just not ready for them.

Interestingly, in the same year that Gurney's stagecoach trip came to grief, 1829, the Manchester-Liverpool railway line had just opened, and was to make its builder, George Stephenson, the most popular man in England. This was five years after the first line, a small one from Stockton to Darlington, had shown the railroad to be practicable.

During a competition to decide who was to build the line, Stephenson's entry got up to thirty-five miles an hour! Spectators were sure that the engine had gone out

* Stagecoach helpers.

of control and that the driver had been killed by the rush-
ing air. But the train was duly brought to a halt, and the
enthusiasm of the spectators was reported to be boundless.

So there was violent opposition from the first to the
steam carriage and yet boundless enthusiasm for the train.
One wonders about the difference in acceptance. Part, of
course, may have had to do with who was involved; those
attending the competition were undoubtedly more used to
machinery than the villagers in Malksham. But there is
more to the question.

I asked René Dubos, who was born in a small village in
France in 1901, whether he remembered the reactions of
his fellow villagers to the introduction of automobiles,
which occurred when he was a small boy. He did, and the
reactions spelled fear.

"Just imagine," he exclaimed, "what it meant at that time
to see automobiles, to have people you didn't know moving
through the village. Before, a stranger was somebody who
lived only five miles away." He hesitated a moment. He
was moving rapidly back through the decades, searching
through his memory to explain the fear. "I remember as
a very young boy—I must have been seven or eight—the
first time the channel was crossed by airplane. It was a
French flier, Bleriot, who did it. Well, that was considered
a fantastic, romantic feat, like going to the moon. So there
was no fear to that. Because it didn't touch us, you see.
It was something that had happened out there, an
extraordinary event. So for this there was only a fantastic
sense of exhilaration.

"But at the same time—I guess it was the same year—
the automobile (not that there were many of them) began
to be seen. I can still remember that if there were crimes
committed in Paris which we saw reported in the newspaper, and an automobile came through, we would all

think we recognized the description of the criminal who was escaping through our village."

The early railroads and Bleriot's flight were a kind of otherworldly occurrence, and were acceptable. It was only when the railroad and even aircraft began to become the "machine in the garden" that opposition grew, and people began to pine, again, for earlier times.

All in all, then, it would seem that future shock, if it exists, is hardly a new malady. And Toffler himself tacitly recognizes this, for he writes, "Western society for the past 300 years has been caught up in a firestorm of change." [29] But if this is so, what is so new about our time? Is the difference just a question of degree?

This is not an unimportant point. Some illnesses are serious and some are not so serious. But a new kind of problem arises when an illness is so new that there is no experience to go on. In such a case the "treatment" is to fumble around until something works or the patient dies. If the patient has had the illness before, however, experience may point the way; after all, he obviously survived the earlier occurrence.

It is also important to determine just how sick the patient is, and whether he is getting better or worse, or holding his own. He may be very sick, but if he's actually getting better, beginning a new treatment might be foolish in the extreme.

The treatment must also separate the physical, mental, and emotional aspects. The patient may have been told over and over again how sick he is, or how bad he looks. He may therefore feel sicker than he is.

I think that is the case with us.

Future shock may be a symptom of our egocentricity—the feeling that everything of importance that ever happened has happened right here and now, or at least within our lifetimes.

Future shock may be a convenient peg on which we are tempted to hang a wide variety of problems—crime, alienation, anxiety, poverty, and many others. But this is too easy; it gives ammunition to the "haves" in their fight to keep things as they are. It permits the nostalgia-prone to fight off all needed change.

But keeping things as they are is not what Toffler had in mind at all. So powerfully, so overwhelmingly has the descriptive aspect of his book taken hold, that the prescriptive part has pretty much been ignored. "Those," he wrote,

> who prate anti-technological nonsense in the name of some vague "human values" need to be asked "which humans?" To deliberately turn back the clock would be to condemn billions to enforced and permanent misery at precisely the moment in history when their liberation is becoming possible. We clearly need not less but more technology." [30]

Basically, he asks for sensible technology sensitively applied. The question is how to do this. Before we can decide on an approach, it will be useful to look into a concept that has come to be called *quality of life*.

2 Technology Has Made a Mess of Our Lives

DON'T CALIFORNICATE OREGON
—*Bumper Sticker*

Has technology created more problems than it has solved? We hear much about dead lakes, chemical poisons, blue-collar blues, destruction of the ozone layer. Computers process us; automobiles paralyze our cities; chemicals invade our food.

A common expression is that "the quality of life is deteriorating." How can we determine this? There is no thermometer, no barometer, no qualitometer that will tell us what we want to know.

Ask

One way to find out whether quality of life is deteriorating is simply to go out and ask people how they feel about it.

The trouble with this approach is that the answer depends on the definition of the term, while the definition depends on the answers that come in.

What is actually being asked in such case is a "gut" feeling. But gut feelings are notoriously dependent on extraneous and even unreliable factors. How is the respondent feeling at the moment? Is it raining or sunny? Has a temperature inversion caused some concentration of pollution, leading to eye or throat irritation? Administrations come and go; harvests are good and bad; the economy runs hot and cold. A question asked one month may elicit totally different answers from the same one asked the next.

Still, this is the easiest and most direct route, and one I have made use of myself. During the two years or so that I have been working on this book, for example, I have held innumerable discussions with friends, colleagues, strangers; students, professors, shopkeepers; all sorts of people. As time went on a very interesting pattern began to emerge. I encountered much pessimism about city life, about our society, about technology, even about civilization itself. But in virtually every case the individual expressed optimism with respect to his or her own future! This included several people who had recently lost their jobs.

Now it is fair to say that my sample is a skewed one, that those I spoke to were mostly members of the middle classes. Nevertheless, many of these people whose personal circumstances and expectations for the future were all quite positive, were decrying the state of our society and the quality of our lives, and even bemoaning the future.

My informal survey meshes well with one that was carried out by Daniel Yankelovich, Inc., and the *New York Times*. New Yorkers were interviewed on a wide range of issues dealing with the quality of life in their city. Almost a third thought that city life would worsen in the next ten to fifteen years. College graduates and higher-income residents tended to hold this belief more strongly than others; but, when questioned about whether they were living in the city voluntarily or out of necessity, these same respondents were more likely to be New Yorkers by preference.[1]

Another study reports that only 35 percent of college graduates were completely satisfied with their neighborhoods, as opposed to 56 percent for those who had an eighth-grade education or less.[2] It seems unlikely that the neighborhoods of the poorer groups were in any way more desirable than the other ones. Clearly, higher expectations were at work among the 35 percent.

But in general the idea, commonly held, that all New Yorkers are beleaguered citizens who pray every night to be released from their trap is much exaggerated. A friend of mine lives in Greenwich Village. We were walking along her street on a fine summer evening and she was telling me about the area. She pointed to a narrow iron gate guarding an alley that led to a house built in a back yard and, giggling, told me that a renting agent, who was trying to interest her in the property, had told her confidentially, "Edna St. Vincent Millay almost lived there!"

Further up the street my friend pointed to a third-story window that had flowers positively spilling out of several lovely window boxes. Just that morning, she told me, she had come out of her apartment, feeling lonely and blue. As she walked down the street, however, she was met with the smiling strains of a Mozart piano sonata; she recognized the touch as that of her friend who lives in the flowery apartment and her mood lifted immediately.

It seems that not every city dweller wants to live in the country. She was living exactly where she wanted to live.

Similarly, I found that even among those of my respondents who were most pessimistic, many were doing exactly what they wanted to be doing. If not, there was almost invariably the hope, expectation, or belief that they would be able to change careers if they wanted to, or could move (or were moving) to another goal.

All in all, it does not seem that asking people whether the quality of life has gone up or down is likely to produce a definitive answer.

Quality of Life Index

Some indicators, being inherently quantitative, are easily measured: gross national product, per capita income, consumer price index, disposable income, years of schooling, age span, amount of sulfur dioxide in the air, square feet of living space per person.

Census figures show, for example, that there is far less of the overcrowding in American cities—five, six, or even more people in a one- or two-room apartment—that was common at the turn of the century. This is fact. There are fewer dwellings without indoor plumbing. This too is fact.

In some cases, like those involving air, water, and land pollution, measurements are possible and do show deterioration. But just because accurate measurement is possible, specific areas where improvement was needed could be mapped and attacked. And we are seeing improvement in certain of the more blatant abuses. In some places missing birds are returning; many rivers and lakes that were assumed to be dead were merely sick and are recovering; [3] the air in some cities is actually cleaner than it has been in years.

These indicators also tell us that we have far to go before we can be said to have cleaned up our environment. They should be used to tell us where work needs to be done—not as a signal to start wringing our hands.

But most quantitative indicators (the great majority of

which have shown improvement over our country's history), are shrugged off. There are other things, we are told, that are more important. Beauty of surroundings, peace of mind, work satisfaction, freedom from fear, and freedom of speech and religion are a few that come to mind. These might be called social indicators, and are clearly just as important as the economic, environmental, and health indicators. It is on the basis of these social factors that the quality of life in our country is said to be deteriorating.

But social indicators, being inherently qualitative and emotion oriented, are obviously more difficult to quantify. We are told, for example, that we are less free now than we used to be. But on what basis? Are we really less free than we were a hundred or two hundred years ago, when slavery and domestication of people were common? Does the word *freedom* mean the same thing today as it did then? Not too long ago the word was likely to refer to freedom from government interference. Now the government is exhorted to "do something!"

How do we balance our public desire for social equality against our private desire for inequality?

And who are "we"? Our country, more than most, is not *a* society; it is dozens, maybe hundreds. Among them are students, blacks, WASPs, Jews, hard-hats, spiritualists, teachers, welfare recipients, doctors, the social elite, the *nouveaux riches*, old folks, civil servants, handicapped, sports fans, music lovers, and so on, each with its own set of desires and needs. And there are federal, state, county, and municipal interests involved as well.

Lewis Mumford claims that machines have cut down on creativity and personal expression—for example, that where people used to sing and hum, they now watch television and listen to records and the radio.

This of course is true, but not the whole truth. These machines also introduce many to music, and develop an appreciation for it that might not have arisen before. Music lessons, guitar and recorder playing, chamber ensembles, school and community orchestras and choruses —all have increased in the years since these inventions were introduced into our culture. (Of course, an increase in leisure time has also contributed to their growth.) Prior to the invention of the phonograph at the end of the nineteenth century, on the other hand, if there was no musician in the home, there was no music.

How can we evaluate, balance, and measure such factors? The answer of course is that we can't—not in a completely satisfactory way. And so what social scientists try to do is choose a set of indicators that *can* be measured, a set felt to be representative of the total, amorphous thing called "quality of life." Then, by rating each of these on some sort of numerical scale for a given area, and adding them up, it should be possible to see, on a numerical scale, just what the quality of life is—here with respect to there, now with respect to then [4]—and perhaps to show with the same clarity as does an environmental indicator just where a society is weak or lopsided and needs work.

The big problem here is to choose the proper indicators, assign the proper weights, and combine them in such a way that it truly is a simulation of the "real thing." Indeed the question is often raised as to whether it is possible for anything as intricate and tangled as a society to be summed up by any sort of workable model, no matter how complicated.

Nevertheless, a number of these representations have already been proposed; indeed, they are springing up like fast-food stands. But there seems to be agreement about them on only one point, that the definitive index has not yet been constructed—indeed, that it may never be.[5]

And even in cases where measurements can be made accurately, factoring them into the equation may not be easy. A big problem today is industrial hazards, particularly chemicals in the air. Recently, it has been shown that vinyl chloride, a widely used substance, is a cancer-producing agent. As a result restrictions on occupational exposure are being tightened (and consumer exposure is being reduced as well).

The dangerous effects of new chemical substances, which are being created at a rapid rate, have been showing up quite regularly. Yet, as is shown in Table 2, occupational airborne particles cause only .5 percent of all chemically linked deaths.[6] The figures were arrived at before the vinyl chloride problem arose, but it is unlikely that they will change much as a result.[7] As can be seen, the major problems by far are the self-imposed dangers of smoking, drinking, and poor food habits.

I do not mean to minimize the problem of occupational diseases, of which one hundred thousand Americans are estimated to die each year.[8] What is most disturbing is that many of these deaths, and the nonfatal illnesses as well, could probably be avoided. For example, it has been known for ten years that asbestos is a cause of lung disease, including cancer, and the substance had been suspected for thirty years before that. Yet industry still tends to play down the dangers; workers and unions tend to ignore them;[9] and, with respect to occupational cancer, the government has only recently (1971) entered the battle. And even here the capacity of the group set up to do the work, the Occupational Safety and Health Administration, is hampered by lack of funds and personnel.

One of the great problems is the difficulty of establishing

TABLE 2

HAZARDS OF CHEMICALS

Factor	% OF CHEMICALLY LINKED DEATHS In 1967 Linked to Factor *
Accidents with chemicals	0.01
Birth control pills	0.01
Suicides involving chemicals	0.25
Occupational airborne particles	0.5
Addictive and narcotic drugs	0.6
Alcoholism	3
Adverse reactions to medication	4
Unknown factors acting as initiators or promoters of cancer	3–8
Poor diet	0–20
Cigarettes	17

* 100 percent = total deaths from all causes = 1,850,000.

direct cause and effect. Cancer is particularly puzzling; about the only thing that cancer researchers have been able to agree on is that it is a multifactorial disease; the cause is not a simple poison, virus or anything like that.[10] Similarly, a United States Court of Appeals decided in 1974 that it could not stop a company from dumping asbestos into Lake Superior, despite the fact that a hundred thousand residents of Minnesota get their drinking water from it, because it could not yet be conclusively shown that any of the residents had gotten ill as a direct result.

Clearly much scientific detective work remains to be done.

So the table is not intended to excuse the occupational situation, which in some areas (geographical and occupational) is scandalous, but it does give a measure of relative danger. Emotional concentration on the dangers—"Technology kills!"—is not helpful; seeing the magnitude of the problem is. It is true that "technology kills," but so does nontechnology. If we manage to keep a level head we are more likely to get stricter laws governing work environments, which are needed, along with better knowledge of what is dangerous and what is not. Certainly any substance that is going to be put into wide use should be tested by the government before many people have been exposed to it.

A New Level

But even if we never come up with a useful quality of life index, attention to the idea is useful. If nothing else, it points to areas that certainly do need work. The New York City subway—dirty, noisy, crowded or empty, bumpy—is often pointed to as an example of what technology can do to people. It has been found, for instance, that the noise level often rises to a point normally reached only alongside an industrial grinding machine. And noise has been shown to cause not only discomfort but tension and, under constant exposure, hearing loss.

But the New York City subway, it should be pointed out, is not "technology": it is bad—or at least outdated—technology. Experience with the subways of various other cities—Montreal, Mexico City, Moscow—will quickly convince the most skeptical that there is nothing inherently wrong with subways. (The Montreal Metro rides on rubber wheels!) Aside from the fact that they can be quiet, beautiful, and fast, they are efficient; in general a single line can carry twenty times as many people per hour as a single lane of automobile expressway.

What we need in other words, is not *no* technology, but better technology! A comment of Arthur C. Clarke's is pertinent here. I once asked him about something in his motion picture *2001* (he wrote the story and screenplay) that puzzled me. At the end we see infancy and old age juxtaposed. Was he trying to show, I asked, a cyclical aspect to life, to evolution?

He shot back immediately, "That's the general idea; but I prefer to think in terms of a helix." As he said this his finger was tracing the shape of an automobile coil spring standing on end.

In other words, there is a kind of coming back to a starting point, but with an extra something added, a something that puts us somehow ahead of, or perhaps higher than, where we were the last time around.*

We need not be locked into all the admittedly unpleasant aspects of mechanization. Nor need the problems necessarily indicate a continuing descent into hell. Let us think in terms of improvement "the next time around."

Consider the matter of bread, or what the food expert James Beard has called "spongy, plasticized, tasteless breads . . . with about as much gastronomic importance as cotton wool." [11]

The truth is that bread, along with so much else, *has* been "plasticized." But, happily, a realization seems to be taking place, a realization that we are losing out, and that bread need not be reminiscent of wallpaper paste. In one month toward the end of 1973 two major publishers issued books on breadmaking,[12] and *Gourmet* Magazine published two articles on the same subject. Much more has been published, though sporadically, on the subject since then. Revolt is in the air.

But often when the revolters realize what it takes to bake a good loaf by hand—including the kneading, which

---

* Giovanni Battista Vico (1668–1744) suggested that progress does not take place in a straight line, but in the form of a spiral. Saint-Simon (1760–1825) too thought in terms of an orderly progression of stable civilizations, each an advance over the previous one.

is *hard work*—their enthusiasm tends to dim rather rapidly. Not all breads require kneading, but they all do require attention, time, and patience.

Some families have taken to baking breads together, making of the process "innings" to substitute now and then for outings. A good friend of mine is a bread baker; he happens to be retired and looking for exercise. All of Murray's friends think his bread is great—but he's gotten tired of supplying free loaves. And we could not possibly pay him the price such bread would realistically be worth. Also, at this point it is still fun for him; as work. . . ?

So, where to from here? The smart bakers will find a way to produce good bread mechanically. There is no reason why it can't be done. It may take longer; it may be more expensive. But it will surely be less costly than doing it by hand (if we figure in the time involved).

To eliminate some of the chemicals in packaged bread, it may also be necessary to shorten and tighten the distribution process, as is done with dairy products. In these days of refrigeration and rapid transportation, this should be no trouble at all.

Then we will have come pretty much full circle, but at another level. And just as something is lost by not doing it yourself, something also is gained: choice.

When the ladies used to gather around the well to draw water, or at the riverside to do their wash by pounding it on the rocks, there was a community feeling, a social hour that was important to all of them. But was it important to them because they had not a great deal else? How many of today's housewives would be interested in trading places with them?

And of those who would—having perhaps just gone through a bout with a leaky washing machine, a vacuum cleaner that spilled all the dirt out again, and a broken toilet—they might find that there are problems in the "simpler" life as well—such things as too much or too little rain, heat, cold, insects, disease, shortages.

Let us not think, therefore, that we have arrived, but rather that we are on our way. If we do not like our present situation, or our present destination, let us choose another. If we do not know where we are going, let us choose a destination. We do have choice.

It may even be that one of the reasons people are angry with technology is that they are being forced to face up to the fact that if our power is really so great in the realm of technology, then our problems must be social, cultural, and economic, not technical. In other words, we are being forced to face up to the fact that it is we the people who are to blame for the ills of our society, and we don't like it. For the truth of the matter is that we probably can do most anything we put our minds, backs, and pocketbooks to—including feeding the poor, educating them, providing jobs for those who want them, and maybe even making that work interesting.

The Industrial World

One of the biggest problems in the world of labor is that of the industrial worker. For years the archetype of this poor creature has been the human bug in the giant machine —tied to its rhythms, dwarfed by its enormous size, dulled by the repetitive character of the work.

Daniel Bell makes a few interesting points here. First, the influence of the large industrial corporation seems to have plateaued. In 1956 incorporated businesses accounted for over 57 percent of the total national income, having increased steadily over the previous decades. Since then, however, the proportion has remained stable.*

Second, he says, we are moving toward a post-industrial economy in which services are more important than manufacturing. And while there are service organizations in banking, insurance, utilities, and retailing that rival the largest industrial organizations, most service firms are smaller, perhaps because there is less need for and dependence on large machinery. And even where unit size tends to be larger, as in schools and hospitals, there is often departmentalization, along with a high degree of autonomy and professional control of the departments.[13]

One should not, Bell adds, "miss the fact that what has been appearing is a multiplicity of diverse types of organization and that the received model we have, that of the large business corporation, while still pre-eminent, is not pervasive. . . . The rhythms are no longer that pervasive. The beat has been broken." [14]

Even within the world of mass production, including that stronghold of industrialization, the automobile industry, there are signs of change.

The Sounds of Discontent

Early in the Industrial Revolution, reform agitators concentrated their fire on the physical working conditions, which were indeed cruel. In recent years we have been hearing about another kind of problem, tedium. "Blue-collar workers are bitter and dissatisfied with their work. . . . In even the most casual conversation, the blue-collar man is openly bitter about his job or organization." [15] The "blue-collar blues" are often cited as a major

---

* Incorporation is of course no guarantee of bigness, but it is an indication.

factor in aggression, drug abuse, alcoholism, delinquency, family instability, and in physical and mental illness in general.

Anecdotal evidence abounds. "Every day I come out of there I feel ripped off," complains one young worker at the highly automated Vega plant in Lordstown, Ohio. "All I feel is glad when it's over . . . I don't even feel useful now. They can replace me. . . . They could always find someone stupider than me to do the job." [16]

Now the gut reaction to this is sympathy for the worker and anger at the system (which of course is what our reaction is supposed to be). But let's reflect for a moment. When did the unskilled worker ever feel "useful"? The poor peasant feels useful mainly because he's feeding his family —if indeed he is. But the assembly-line worker is doing a better job of this than the peasant could ever hope to do. Another implication is that the Lordstown worker quoted above is smart enough to do something "better." So? Let him go to school and learn to do something else. That plant and many others have tuition refund plans.

Again, we face the question of whether contemporary complaints stem from truly repressive conditions or from increasing expectations of a decent life. Anatole Broyard of the *New York Times*, in reviewing a book called *Working*, points out that

> many workers seem to feel that their employers owe them not only a job, a salary, insurance and a pension plan, but a philosophy as well. They want to know "why" they are doing what the do, and perhaps it is time every union had a Zen master to answer this question for them. It certainly doesn't strike me as the employer's obligation. He is lost in that very same limbo.[17]

But there is an even more basic phenomenon at work here. The blue-collar blues have been so self-evident that there has been little incentive to study the matter. This has begun to change. To everyone's surprise, several studies have shown that factory workers are by and large contented with their work and working conditions! [18]

One recently published study carried out by four members of the Rutgers Medical School Department of Psychiatry, indicates that 95 percent of the workers at a General Motors assembly plant in Baltimore are satisfied with their jobs. Seventy-one percent maintained that they did not find any part of their work upsetting or tiresome.[19]

It would seem then that much of the "dissatisfaction" we hear and read about comes from *non*-blue-collar workers, from scholars and writers who are telling us what they think the workers should be feeling, perhaps what they themselves feel. And, of course, the discontented minority of blue-collar workers can make plenty of discontented sounds.

Making Work Interesting

Is work in a plant *inherently* dull and repetitive? Arthur S. Weinberg, coordinator of the Worker Exchange Program, New York State School of Industrial and Labor Relations, tells of a visit to an automobile plant in Sweden:

> In the engine assembly division, workers are grouped in multiples of three per engine assembly table. They can choose to assemble it as a team or assemble it individually. Most of the machinery had been designed by Saab-Scania and was later purchased by various international machine manufacturers. The noise level of the machinery was far below the decibel

level of comparable American machines. . . . In contrast to the noisy and dirty conditions of comparable American plants, one could not help but be astonished.

Outside of the plant was a beautiful lake owned by the company. Workers from the plant were sailing, playing tennis on the grounds, and enjoying a beautiful location in one of the garden spots of Sweden. I saw no evidence of industrial pollution but rather felt more as though I was in a protected national park.

We have much to learn from our Swedish corporate counterparts. . . .[20]

The Volvo Company too, especially in their new Kalmar plant, is reported to have gotten around some of the worst of the boredom associated with assembly-line work by setting up teams of workers who divide up the work among them in such a way that each person does not have to do a single, boring job.

Unfortunately the direct costs of manufacturing in this way are higher. Volvo's new Kalmar plant is estimated to have cost $2.5 million more to build and equip than a conventional automobile assembly plant Labor costs too tend to be higher. But the Volvo and Saab are relatively expensive cars, so the higher costs can be absorbed somewhat more easily than in low-priced American cars.

On the other hand, the higher costs of manufactured goods might make us cut down on their use and hence save some raw materials. And if we built our cars to last longer, this too would cut down on use of raw materials.

In Norway, nearly half the thousand workers at the Hunsfos paper mill are taking part in experiments with autonomous groups of workers. Each employee learns and does several jobs. The groups work without a foreman, maintain their own quality control, set their own vacation schedules, and even order materials. Management is called on only for advice.

The changeover has not been easy, however. Many workers are suspicious that management is in some cunning way trying to increase the work load or cut the work force. Others know what they have and fear any kind of change. One of the disadvantages of the program at Hunsfos is the large amount of time required of management to keep the program going.

A growing number of American companies are working on what have come to be called *job enrichment* or *job enlargement* programs. Of course, one approach to the problem of the boring job is simply to eliminate all such jobs by means of more automation. For the truth is that job enlargement programs, even the more advanced ones, by no means create a utopian situation. A certain amount of tedium is inevitable due to the very nature of the work; that is, efficient manufacturing is based on elimination or reduction of variation, while in a work situation it is variation that provides interest. The same holds, of course, for white collar work as well.

Even if we had a labor shortage, which we certainly do not at this time, there would still be problems with the idea of instituting more automation. One is that automated machinery is expensive. But more important, it would not only remove jobs from the market, it would remove unskilled jobs, which are the main port of entry into the labor market for the unskilled young. The obvious answer, theoretically, is to make sure everyone has some skill and a place to use it. While this is clearly an impossible dream, that doesn't mean we shouldn't try for it.

Under the present circumstances it would seem better to make the job more interesting than to eliminate it. Perhaps what we need is some sort of compromise, in which more automation is used not to speed up production but to eliminate the most tedious work.

The Young

It appears to be mostly the young workers who, for various reasons, are doing the complaining. This is especially true in the highly automated automobile plants, where job specialization has reached what must be close to its ultimate. One reason, surely, is their having become used to a generally higher level of material comfort and having gotten more education than their parents, which makes them less willing to put up with boring, repetitious jobs.

Margaret Mead, René Dubos, Charles Reich, and others give credit to the younger generation for their unwilling-ness to put up with such boredom, and, among some of the better-off youngsters, for their jaundiced view of materialism and their willingness to adopt a simpler life style.

But maybe we of the older generation should take some of the credit for having somehow gotten across to the younger generation the idea that materialism is not the greatest and most important thing in the world.

But for those who still feel that the quality of life has actually deteriorated in recent times, and that the only way out is the way back, let's take a look at what life was like in earlier times.

3 G.O.D., or The Good Old Days

The 'good old times'—all times when old are good—
Are gone.
> —*Lord Byron, "The Age of Bronze" (1823)*

The golf links lie so near the mill
  That almost every day
The laboring children can look out
  And watch the men at play.
> —*Sarah Norcliffe Cleghorn, "Quatrain" (1915)*

Don Quixote, that great creation of Cervantes, is often pointed to as the epitome of the brave soul who recognizes and fights against the crassness of modern civilization. In his clash with the windmill, he is recognized as the standardbearer for all technophobes to come.

Indeed modern technophobes are also tilting at windmills, and their view of modern technology is almost as distorted as was Don Quixote's view of the windmill. For it is based, largely, on a false foundation—the idea that what existed in the past was better than what we have now.

The question is, Better for whom? Those who take this stand are typically the educated, the opinion leaders—in other words, the wealthier or at least better-educated citizens of our society. They therefore identify with the higher-class citizens of yesterday. And the higher-class citizens of yesterday did indeed have it very good.

After all, technological innovations are often labor-saving devices: washing machines, sewing machines, vacuum cleaners. But the upper classes never had to do the labor that these devices are saving! They had servants, or slaves, to do all of this. So to them labor-saving devices have not been nearly so important as they have to the middle and lower classes.

At the same time we are conditioned, by the nostalgia industry, to think of the past in glowing terms. Were things really better for the average person in days gone by? *

Were people better off half a century ago? Buckminister Fuller says that in his first jobs before World War I, he found "all the working men to have vocabularies of no more than 100 words, more than 50 per cent of which were profane or obscene. . . . Their intellects were there, but dulled and deprived."[1]

The turn of the century? Overcrowding, hunger, desperate poverty existed then, as they do now. The main difference is that they were suffered by a much higher percentage of the population, yet little attention was paid to their plight. The rich were very rich and the poor were hardly noticed.[2]

The major troublemaker of our time is still thought to be the automobile. Wouldn't it be just great if we could

---

* The following chamber of horrors will annoy some readers and anger others; it also leaves me open to the claim that I am deliberately choosing to report the negative side of the past and ignoring the positive side. To this I can only answer that (1) the nostalgia industry takes care of the positive side; (2) I have tried to include broad categories of negativity, hallmarks of the age, as it were, not single or personal hardships that have existed through all time; and (3) we need to be constantly reminded that we have progressed *as a whole*, not retrogressed, if only to counter the nostalgia lobby.

get rid of it and go back to the horse and buggy? Clean air, clean streets, free fertilizer. What could be better, more sensible? But before it's collected, manure breeds flies like flies, and these carry some thirty diseases that they spread wherever they go. A small city might generate 130 tons of manure a day, to be collected and carted away; a city like New York, ten times as much. Some of it was used as fertilizer, of course, but the rest was simply dumped in a convenient spot. New York dumped it on an island in the Hudson River. Urine in the streets made them slippery for horses as well as humans. If a horse fell and injured itself badly enough, it had to be done away with. In a city like New York as many as fifteen thousand horses had to be hauled away each year, and if you think a dead car on the streets is unpleasant . . . The horses' footing was better on cobbled or brick streets, but then the filth collected in the cracks. Runaway horses and toppling coaches were all too common.[3]

Further Back

How about a century ago? Surely that was a better time than now? Think of the condition of our prisoners, our mentally retarded. Indeed, we think of it fairly often. Today we find conditions ranging from unpleasant to cruel. But we call them news. A hundred years ago it was quite common for the mentally retarded and the insane to be locked away, often in chains, and simply forgotten about.

   Things may not have improved much since then, but they *have* improved. We rarely lock the retarded in chains now; nor do we beat people for what are now recognized as symptoms of Huntington's chorea (involuntary, jerky movements of the body and limbs), trying to drive the devil out of them.

We often hear complaints now that we are too permissive, and perhaps we are. But somehow that seems an improvement on flogging, cutting off a hand, or execution as a punishment for theft. Indeed, two hundred years ago a man could be hanged in England for any one of two hundred offenses, including stealing five shillings from a store.[4] And in an era when there was little relief, or welfare, and no social security, there was often good reason for stealing.

On November 20, 1878, for example, a cable dispatch was sent from London that read: "A state of appalling distress and destitution exists among the mechanics and laborers of Sheffield . . . living in tenements without clothing or furniture, which they have been forced to sell to procure food . . . dependent upon the charity of their neighbors for subsistence."[5]

You might read, if you have not already done so, accounts of the life of the common man in that day—life aboard ship, as in Melville's *Billy Budd* and *Moby Dick*, or in the industrial areas, as in Dickens's *Hard Times* and *Bleak House*, and Rebecca Harding Davis's *Life in the Iron Mills*. Thomas Hardy's *Jude the Obscure* shows clearly that life in a small town can be quite as stultifying as that in a large city.

Little equivalent literature about earlier farm life seems to have come down to us. There is a kind of "farm filter" effect; anything that speaks against the delights of farm life is suspect. It also seems reasonable that poor, illiterate farm families were not very likely to produce great writers. And great writers, by definition, are those whose works live on.

But some literature exists; see for example (if you can find them) *The Story of a Country Town* (1883) by Edgar Watson Howe, and *A Son of the Middle Border* (1918) and *Main-Travelled Roads* (1891), both by Hamlin Garland.

It was not until 1852 that a compulsory education law was enacted in the United States. That was in Massachusetts, and required children from the ages of eight to fourteen to attend school for twelve weeks a year. Southern states did not get this far until the first two decades of the twentieth century!

We do not usually think of our grandparents and great-grandparents as particularly cruel people. But who among us could conceive of sending five- and six-year-old children down into mines, often to sit for twelve hours in pitch darkness opening and closing a trap door? Of harnessing human beings to coal cars and making them pull these on their hands and knees? Our grandfathers both accepted (countenanced?) it and did it. The physical deformities you sometimes see in paintings and drawings of earlier ages were not always caused by disease.

In 1850 the average work week in factory, office, or store was seventy-two hours—twelve hours a day, six days a week. And this is what people left the farms *for*. Men, wrote Augustus Jessopp in 1892, "do run away from the odious thought of living and dying in a squalid hovel with a clay floor. . . ." [6]

In Europe, large numbers of the lower classes, and the middle classes as well, were employed by the better off as domestics. Although they were usually better fed and better housed than the common workers, their lives were, in a very real sense, owned by their employers, and they were thought of as another breed. It was not until 1879 that the lower classes were permitted entry to the British Museum.

In France today a chief justice might earn four or five times as much as his office boy. In 1800 he earned fifty times as much.[7]

Things were so good in the good old days that from the year 1820, when records of immigration first began to be kept, to 1930, some 35 million immigrants came to these shores from Europe, daring to hope for a better life. There are two possible conclusions: (1) life must have been pretty bad; and (2) economic improvement *is* important, and not only to Americans.

The immigrant farmers came to the United States expecting the Garden of Eden and got the wilderness—a bleak, back-breaking existence that only the most incredible fortitude and a well-cultured stoicism enabled them to survive.

Almost any attempt by workers to organize or act together in any way to improve conditions was termed "conspiracy," a device the government used, as Clarence Darrow pointed out, to punish the crime of thought.

Lest you think that these hardships are a product of our industrial civilization, listen to the words of anthropologist Margaret Mead. "Through the centuries," she writes,

> most of the peoples of the world have lived close to fear—fear of hunger, fear of cold, of chronic illness, of ignorance. In those societies or at those periods which later have been called *great*, a small proportion of the population have been elevated above some of these fears; their food and drink, the care and protection of their children, their control over the knowledge that mankind had accumulated so far was assured. The others, ninety-nine percent, remained relatively wretched. . . .[8]

During the years 1877–78, for instance, in North China,

when floods destroyed food stocks over a large area, some 9 million inhabitants are estimated to have starved to death.

Perhaps we can find Eden in the eighteenth century? The naturalist Carl Linnaeus wrote about a famine in Sweden:

> I fear that I shall not have any under-gardeners this summer to do daily work, for they say they cannot work without food, and for many days they have not tasted a crust of bread. One or two widows here are said not to have had any bread for themselves or their children for eight days, and are ashamed to beg. Today a wife was sent to the castle for having cut her own child's throat, having had no food to give it, that it might not pine away in hunger and tears." [9]

It has been calculated that in the first half of the eighteenth century 75 percent of all children born alive had died before their sixth year.[10]

An average Frenchman in the seventeenth century, according to a UNESCO study, could expect to lose one of his parents by the age of fifteen, and two or three of his sisters and brothers before he got married; if he happened to reach the age of fifty he would probably have lived through two or three major epidemics, two or three famines,* and three or four near famines.[11] (The Hasidic leader Menahem Mendel once said, "It is not hunger I fear; it is the cruelty that comes from hunger that frightens me." [12])

The life of a married woman in England at that time was not very different from that of a slave. Garrett Hardin

---

* There are still famines today, it is true, but they do not occur in the industrialized countries.

points out that she (1) could not vote; (2) could not own property of her own; (3) could not keep any income of her own out of her husband's hands; (4) could not divorce her husband even for repeated and flagrant infidelity, though one misstep on her part was enough to get her husband a divorce; (5) could be completely cut out of her husband's will (in one notorious case, the small fortune a wife earned as a milliner while her husband was unemployed was willed by her husband to his illegitimate children, leaving the wife penniless at the age of sixty-two); (6) could never refuse a husband's advances, though he was under no obligation to her; (7) had no legal complaint if her husband beat her.[13]

Yet we still hear cries about what we have lost. Jacques Ellul, one of our prime antitechnologists, tells us that "the peasant commune and the peasant family were slowly ruined in the eighteenth century. The historian notes the collapse, relentless and more rapid [in England] than in France, of a whole society which had been in equilibrium until then."[14]

Equilibrium, balance, stability. Everyone knowing his place.

But if your place happened to be that of a slave or a serf or a starving peasant, I wonder how the idea of a stable society would seem. Again, it would seem the antitechnologists view the *older* societies from the top.

And again, people writing in those good old days did not seem to think things were so beautiful. Jean Jacques Rousseau wrote in *The Social Contract* (1762), "Man is born free; and everywhere he is in chains."

And just as there has been a recurring romantic movement throughout history—a romantic movement among whites—romanticizing the past, so too has there been a black romantic movement. J. K. Obatala, of the black studies department at California State University, tells us:

> Were the average African-American familiar with the rigid social stratification of Egyptian and other ancient black societies, there would hardly be for him anything romantic about them, since their ways of life were based upon the drudgery and ignorance of the common man. Still less appealing would be the tyranny of the long Islamic oppression in sub-Saharan Africa, which has been romanticized by certain Westernized African and African-American intellectuals.[15]

Witchcraft and Dirt

From the end of the fifteenth century, when the Inquisition took charge of the "job" of exterminating witches, until 1782, when the last official victim was killed, several hundred thousand victims are believed to have perished in Europe alone. In some countries, especially Germany, so many victims were taken that villages were virtually depopulated, and in some cities commerce came close to collapse. In the year 1275, not far from the city of Troyes, a great throng of spectators watched 183 people being killed in a mass burning.

This, of course, was an aberrant aspect of religion which is not directly tied into technophobia. There is, however, an aspect of religion that is. For what we may be seeing today, and in many earlier recurrences as well, is a hangover from the old Calvinist/Lutheran/Puritan frame of mind, in which one honors the Creator best by belittling His most worthy creation. The more despicable man is made to feel the better; guilt feelings are good.

These feelings may derive from an even deeper and

older fear, a fear of the gods that can be assuaged only by saying aloud, so that the gods can hear, that what such believers possess is as nothing, that their accomplishments are worthless, that their strength is minuscule, and their worth less than dirt. Anyone who says, and even dares to believe, otherwise, tempts the wrath of the gods.[16]

But if their worth was less than dirt, then the people living in the Middle Ages lived at the right time, for there was plenty of it.

Until the widespread introduction of cotton into Europe about the fourteenth or fifteenth century, for instance, wool and to a small extent leather were the only materials available to the masses as protection against the cold and drafty buildings of the time. (The period from about the fifteenth to the mid-nineteenth century is sometimes referred to as the "little ice age.")

Natural wool is washable only with difficulty, however, while leather is not washable at all. Dry cleaning is a relatively modern invention, dating back to about the middle of the nineteenth century. Thus grease stains in the past were cumulative; garments accumulated them like mementos of days gone by, and became dark and shiny from grease, oil, and wax. The luxurious satins and velvets of lords and ladies were virtually impossible to wash, and became and remained stained and spotted as they were worn.

Even plain soap was expensive, hard to come by, or simply ignored because of water problems. It is said that Queen Isabella was bathed only three times in her lifetime—and only once voluntarily. That was when she married; the other two times were at her birth and death.

Surely that is exaggeration? Perhaps. But as recently as the mid-1800s Macaulay wrote in his *History of England:* "The ambassador of Russia and the grandees who accompanied him were so gorgeous that all London crowded to stare at them, and so filthy that nobody dared to touch them. They came to the court balls dropping pearls and vermin."

The main advantage that the rich had over the poor was that they could afford to change their garments, and could purchase new ones, rather more often—and, if they were interested in soap and hot water, could afford it.

Thus it can be said that technology has been one of the greatest aids in egalitarianism by providing rich and poor alike with low-cost means of staying clean. If you're dirty and disgusting today, it's your own fault. But in the old days the poor were not called "the great unwashed" for nothing.

Nor was this level of cleanliness restricted to the person. In medieval Europe the alleys were public lavatories, the doorways sink drains, and the windows garbage chutes.

Due to poor diet, teeth were a chronic problem, not so much from cavities as from diseases of the gums. And with rough dirty wool worn next to the skin, skin diseases were common.[17]

Thersites, in Shakespeare's *Troilus and Cressida*, gives us a listing of some of the commoner afflictions of his era.* He speaks of "the rotten diseases of the mouth, the guts griping, ruptures, catarrhs [running noses, etc.], loads o' gravel i' the back, lethargies, cold palsies [shakes], raw eyes, dirt-rotten livers, wheezing lungs, bladders full of imposthume [festering swelling or cyst], sciaticas, lime-kilns i' the palm, incurable bone-ache, and the rivalled fee-simple of the tetter [pustular eruption of the skin] . . ."

Insects, rats, and other disease-carrying pests of all kinds were common. When these were combined with the almost

_____

* Though the play is set in Roman times, it is likely that the listing reflects Shakespeare's era, or the end of the 1500s.

total lack of sanitary conditions as we know them, the result was malaria, typhus, dysentery, and other such diseases. And when bubonic plague struck (transmitted by fleas from infected rats), it simply ran amuck. One estimate is that within twenty years, starting in 1334, as much as three-quarters of the population of Asia and Europe may have died.

The Age of Drabness

Such are the more obvious disadvantages of life in earlier times. But technology has brightened our lives in other ways that we never think about. Consider, for instance, what life would be like if all your clothing and all fabrics about you were drab gray or brown, which they were for the poor. Marc Seguin wrote in 1839 with wonderment of how

> after being submitted to a chemical preparation for some hours these clothes dipped into a vat came out as if by enchantment with the most lovely colors, with the most graceful designs. These make the pretty prints, which on holiday, the working class itself wears, and which in the country as much as the town, light up with their sparkle and freshness groups of the young girls and spread all around them an atmosphere of joy, comfort and happiness.[18]

Prior to the early nineteenth century, which roughly marked the beginning of the chemical dye industry, all dyes were natural, derived from vegetable matter, minerals, and even insects and shellfish. Obtaining the color by natural methods tended to be a time-consuming and complicated process, so dye materials were expensive.

Today as part of a larger back-to-nature movement, we

hear again about colors from nature. But living in a world of color, in a time of relative ease, in an economy of plenty, one has the energy to engage in natural dyeing; it's fun. In earlier days there was likely to be little inclination for such activities, or enough energy left for them after a day's work.

For the poor, then, the world was a drab place indeed.

The Golden Age

Well, then, how about the Golden Age of Athens? The first democracy! Surely that must have been an exciting, a desirable, time in which to live. Perhaps it was, if you happened to be one of the full citizens. But the odds were against you. Fully half the population was enslaved. And of the remaining group, half were women, who were in general considered useful only for childbearing. Ideal love was that between man and man.

Only the full citizens set the tone of that admittedly incredible city-state. Only they produced the marvelous arts and sciences and philosophies for which the time is justly famous. But then, they didn't have to spend their time and energies on the toil of daily life. Their daily needs were taken care of by others who were not so favored.

In other words they, like our own plantation owners two millennia later, and like many others in other places and times, lived the good life by sitting astride the backs of the less fortunate. It happens today too, of course, in an economic sense. But the differences are truly enormous. Today one must be very rich indeed to be able to afford live-in domestic help. And in some areas domestics are organizing their own labor unions.

Violence

We keep hearing that we now live in an age of violence. But the very greatness of Athens seems to have been born of violence, namely the Persian Wars (500–449 B.C.), and to have ended in the wars with Sparta, Macedonia, and Rome. The Roman era was a time of continual invasion, conquest, and extermination.[19]

Superstition played its part as well. A thousand years after the Roman era the Vikings of ancient Norway "bought" the safety of their ships by binding sacrificial humans to the rollers over which the ship was to be launched. The blood from the broken bodies provided the gods with drink.

In recent years we have been stunned by the assassinations of John F. and Robert Kennedy, the killing of Martin Luther King and the attempt on the life of George Wallace. But rulers throughout history have been the target of dissidents or religious fanatics. The very word *assassin* comes from a twelfth-century Islamic order whose members considered eliminating the leader's foes a sacred duty.

Professor Leo Hershkowitz, director of the Queens College Historical Documents Collection, in comparing the violence of today with that of New York in 1800, maintains that those who complain about life today "wouldn't have lasted a week then." [20]

Even the perception of violence varies with culture. F. M. Esfandiary, author and teacher, tells of a discussion with a Middle Eastern friend about violence in the latter's country. His friend maintains there is very little violence there. Esfandiary challenges him:

"I see parents beating their children."

"That isn't violence," he said. "They want to bring up their children properly."

"How about corporal punishment in your schools?"

"That too isn't violence. It is part of education."

"Just recently another clan war erupted in one of the villages. How do you explain that?"

He smiled and shrugged his shoulders. "Those are just family quarrels."

"Then what is violence?"

"Violence is what you find in the streets of New York —muggings, rape, juvenile delinquency." [21]

The Good Old Days Today

It is still possible, if you are so inclined, to experience the good old days. There are many places in the world where life still goes on as it used to, and there are even places in the United States where this is more or less so. There is still poverty and malnutrition in the United States today; whether it has been caused by our industrialization is debatable; other industrialized countries have managed to do away with these scourges, and we could too if we had the will, but perhaps at the expense of some of the freedoms and luxuries we have become used to.

In the less-developed countries, however, it is the majority who are poor, who are suffering and struggling. Indeed, a billion people in the world today are said to be malnourished, most of them concentrated in some thirty-two countries spread across the tropical areas of the globe. A terrible drought has ravaged sixteen nations in Africa the last couple of years; already a million inhabitants have died of starvation and disease.

In Chad the undernourished children were unable to resist an outbreak of diphtheria, but the leaders requested that drugs *not* be sent in: diphtheria, they said, was at least a quicker and more merciful death than starvation.

Scourges like spinal meningitis still sweep across some parts of the less developed world, while diseases like schistosomiasis—diseases that leave the victims only half alive—are endemic in tropical areas like Egypt.

In India the bondage of peasants to a landlord, though illegal, still persists. And the Masai, those handsome cattle herders, still perform clitorectomies on the young women to prevent promiscuity.[22]

And as for the island paradises of Southeast Asia, you can rest assured that they too have had their share of disease, massacres, and tribal warfare.

But the Tasaday. Yes, there are the recently discovered Tasaday, whose gentleness and loving ways are a joy to behold. A tribe of twenty-six people right out of the Stone Age, living in a land that supplies all their earthly needs.

Shall we all go there? How long would they last? Ten minutes? Already the Philippine government has erected an imaginary wall around them, trying to protect them from the incursions of the outside world.

To blame technology and the industrial revolution for the world's ills, then, is to ignore history, to ignore the roiling, steaming anger and misery that built and built and finally erupted into the various revolutions in history— the Reformation, the French and American revolutions, the Civil War; into the Hundred Years' War, the Crusades, the Great Emigration, even Christianity itself.

The Noble Farmer

In spite of all this, there yet remains, safely locked up in the breast of everyone who faces a long line of traffic, or looks up at a cloudless sky at night and sees not a single star, the feeling that the rural way of life is somehow "right," while city living or modern civilization is somehow "wrong." That cities without pollution and congestion are possible is ignored. Always there is the vision of the happy farmer—smiling, proud, dignified, at peace, as contrasted with the tense, uptight organization man or factory worker of today.

But Guy E. Swanson, of the University of California's Department of Sociology and the Institute of Human Development, points out: "A standard questionnaire on psychosomatic symptoms shows no differences among respondents by their degree of modernization." [23]

City people have a curiously bifurcate feeling about country people. On the one hand there is contempt, real down-to-earth contempt—the common caricature being one of the bumptious, gangling farmer, a rube, a hick, with the straw hanging from his lips. Yet, on the other hand, there is the deeply rooted belief that the farmer really lives a much more beautiful, peaceful, pleasant life. The very words *simple*, *rustic*, *bucolic*, *agrarian*, and *Arcadian* seem to conjure up images of beauty, joy, honesty, openness, and all the other positive pleasures and values we can think of. There is a suggestion that such a life is somehow natural, while city life is unreal, or unnatural. Not everyone falls into this trap, of course. Rousseau's vision of the idyllic life of man in a state of nature was so beautiful that Voltaire, after reading it, wrote to Rousseau that it filled him with a longing to go on all fours.*

The reaction of John Fiske, the historian, was a little more direct. He called Rousseau's Noble Savage an in-

---

* Rousseau's basic idea is that of man being naturally good, but corrupted by society.

sufferable creature whom any real savage would justly loathe and despise.[24]

I wonder, then, if those who yearn so piteously for the rural life of old are not comparing today's industrial society with *today's* rural society—which benefits enormously from the technology of the first.

The painter Andrew Wyeth, for instance, with his love for rural Americana, lives in Maine in the summer and on a farm in Pennsylvania in the winter. Does he get about in a horse and buggy as one might reasonably expect? No, he uses a $38,500 Stutz Blackhawk. Custom made in Italy for him, it contains a Cadillac engine and custom-fitted luggage. [25] Thomas Jefferson strongly espoused the farmer and farm life, but was himself interested in good books, expensive wines, gourmet foods, and other such trappings of urban civilization.

Margaret Mead has studied many cultures; she maintains that peasant life per se is not a beautiful and satisfying way of life, that it can be cruel, sordid, and unhealthful, a poor device for allowing its members to rise to their full potential.

Prof. Paul Shepard, in a fascinating book called *The Tender Carnivore and the Sacred Game,* goes even further. His feeling is that farm life tends to create or engender a crude mixture of rectitude and heaviness, an absence of humor. He says that "peasant existence is the dullest life man ever lived. Since the beginning of the modern world the brighter and more sensitive children have left the dawn-to-dusk toil of the farms to their duller brothers and sisters." Yet, he adds, "how tenacious is the illusion that there is something gracious about a life of manure-shoveling, sluice-reaming and goat tending." [26]

In his article, "How Would You Like to Be a Peasant?" Joseph Lopreato quotes a peasant from southern Italy who emigrated to Canada and then returned to visit his parents:

> The village is too small a world to live in. It is impossible to breathe freely in it. It is dirty; you must always hide something or from someone; everyone lies about everything: wealth, eating, friendship, love, God. You are always under the eyes of someone who scrutinizes you, judges you, envies you, spies on you, throws curses against you, but smiles his ugly, toothless mouth out whenever he sees you.[27]

Public opinion surveys taken in southern Italy, Lopreato reports, indicate that the Italian peasants do not wish their children to be like them. One survey of the entire country is particularly revealing. It indicates that 25 percent of the people in the industrial north would have liked to see their male children become farmers, but only 8 percent of those in the agricultural south felt the same way!

Lopreato adds:

> Surely there is nothing in the economic history of south Italy that suggests less relative hardship in the past for the peasantry. Indeed there is every indication that the opposite is true . . . [But the peasant] now has a greater awareness of his hardship and suffers from a deep sense of relative deprivation.[28]

Relative deprivation. It seems that this is the *punctum* of the question. If they *feel* deprived, when they compare their civilization to ours, who are we to say that they are not deprived?

What else is the present discontent in our own land but a manifestation of relative deprivation? What is the significance of the growing gap between black and white, between less-developed countries and developed countries, if it is not this? Was the slave, who had no hope of change, better off than the black of today, whose very anger reflects the possibility of change?

Those who point to the earlier, more exciting, "freer" days of the hunter and fisher (Professor Shepard, for example) ignore, in the usual male chauvinist way, an important fact. And that is that there is a certain amount of drudgery involved in running any society. The men in hunting societies may well have lived more adventurous and freer lives than their male descendants do today; but they passed the drudgery of food preparation and cooking, making clothes, tanning hides, and so on *to the women*.

This industrialized society of ours—this mixed-up, technological, nonaristocratic, scientific, "rational," "efficient" society of ours—has given a wide class of people, including the poor, the minorities, and women the idea that they are actually human beings; that they are *entitled* not only to adequate food, clothing, shelter, and health care, but to educations as well; not only to jobs but to *interesting* jobs. And dignity! Desires for job satisfaction and a higher quality of life are modern concepts.

Why this should have happened in the most mechanized, rationalized, industrialized, technologized time in history is really a puzzle. It almost makes one think there might be a connection.

Social critics tell us constantly that the material satisfactions provided by technology have not added to the significance of life or created real happiness. That's like complaining about a plow because it doesn't sing us to sleep. What science and technology have done is to raise our sights (expectations, if you will), as well as to *make it possible* for a large percentage of us to live decently for the first time in history. It's up to the rest of society to put that ability to work, to make that dream come true.

The early utopists felt that if a state of grace was to be found, it would be in the future. In this they seemed to have more sense than the current crop, who keep looking back forlornly, as if we had somehow passed it by without realizing it.

4 *Then Why the Technophobe?*

Urk, accustomed to eating his meat raw, stumbles upon
Swak and sees him roasting his meat over a fire. Having
been burned in a forest fire some years earlier, Urk glares
at Swak and says nervously, "You never were satisfied with
anything, even as a kid. Always trying to 'improve' things.
Don't you realize how dangerous fire is? You'll kill us all.
Why don't you let well enough alone?"

Fear is an emotional reaction to a real or imagined
danger. The reaction can range all the way from a vague
feeling of uneasiness to actual pain.

The case of my own daughter may prove instructive.
My wife and I never heard her cry during the first couple
of days after she was born, and we began to fear that there
was something wrong with her vocal apparatus. We said
"Boo!" and waved menacingly at her, but she remained
quite placid. My wife finally voiced her fears to the doctor.
He pinched the baby's bottom, and she let out a most satisfying wail. Our "fears" were allayed.

But it isn't always that easy.

Fear can be such a strong emotion that it can literally

take over one's life. For such a person just going out to the store can be a terrifying experience. We all know of people who are afraid of flying, of elevators, or open or closed spaces; of blood, fainting, or hospitals; of a myriad of other causes. In such cases the term *phobia* is often a more accurate one.

In fear there is a specific external danger, and the response to it is proportional to the danger. A phobia is also an emotional response to some specific cause, but in this case a situation that would cause no alarm in the average person has been magnified out of all proportion to its actual danger.

At the beginning of a true fear response to an immediate danger, the body prepares for fight or flight. The adrenal glands pour adrenalin into the bloodstream. This speeds the flow of blood while at the same time decreasing its clotting time; it raises the blood-sugar level, oxygenates muscles, and delays the onset of fatigue. Thus under appropriate circumstances the fear response is useful and even, at times, lifesaving.

But whereas in a normal fear response a person may be more alert to what is going on around him, in a phobic reaction he is also concerned with the physical response that is taking place inside his body. Instead of handling the situation or stimulus, his main objective is to flee, to get back to "safe" territory.

And, mainly, his fear is now aroused by an object or event that has only a symbolic connection with what actually frightened him. The sound of thunder, for instance, may bring on a state of panic in a child or adult, because it is the symbol of his father's voice.

In other words a phobia, like a fear response, is an adaptive mechanism; it serves a purpose. The purpose of phobia is displacement. Instead of carting a feeling of anxiety around all the time, the phobic person substitutes a specific fear instead (though not consciously, of course). Thus, if he can keep away from the specific object or phenomenon, all is—or seems to be—well.

The physical reactions involved in phobias can be truly extreme—up to and including fainting, heart attack, and even death. Since the body is not equipped to maintain itself in a perpetual, or long-lived, state of expectancy or "fear preparation," the phobic person may find himself weak, shaking, and confused.

An opera singer with long experience went to bed regularly for four days before every performance. He had a deathly fear of getting laryngitis, though *it had never happened.* Nevertheless, each time his sickness was real enough that the opera management wondered whether the performance should be canceled.[1]

A man I know has had a deathly fear of elevators for years. He himself feels that if he understood how they worked, it would help alleviate the fear. But if someone tries to explain the mechanism to him—even to explain the safety features—his mind closes up like a clam. "Somehow," he sighs, "that stuff never made sense to me."

A similar situation obtains even in something as "dangerous" as flying. While it is certainly true that an accident can occur in flight, the passenger is actually some twenty times safer in a commercial airliner than he is in his own car.[2] Yet he or she will drive at sixty miles an hour after a couple of drinks without a second thought, while getting on a plane may *require* getting potted.

Another basic cause of fears, then, is unfamiliarity. There are very few people today who are deathly afraid of cars, though there is good reason to be, because cars have become part of our lives, while there are still a surprisingly large number of people who even now have never flown.

In other words, there is a fear of anything that is strange. Eric Hoffer, the longshoreman-philosopher-author, tells of working on a farm and having real feelings of fear when told he was going to be switched from picking peas to picking beans! [3]

If, then, unfamiliarity breeds fear, we can begin to understand why even so worldly a person as Miguel Angel Asturias, Guatemalan novelist, poet and diplomat, can write that "a novelist or writer like myself looks with timid respect on everything relating to science, scarcely daring to inquire into, to glance at, the awesome discoveries of the scientists." [4]

Part of the problem may lie in the jargon that is used in science and technology, as it is in any profession. Jargon may be used to bamboozle or impress; it may be a true shorthand; or sometimes, because an idea is really new, current words or expressions are simply inadequate. The result is the same; and the nontechnical person may feel locked into a system for which he holds no key—neither the key to understanding nor the key to getting out of the awesome system.

Still, things are not really as bad as some people make them out to be. When, for example (and if) we develop computers that are really smarter than people, will we—as some people predict we will—bow down to them? Well, consider the steam shovel, which is much more powerful than any man in a physical sense. When was the last time you or anyone you know bowed down to it, in reverence or fear, as a primitive man might still do? Clearly, it is a case of technological sophistication and a sensible relationship between the man and his machine.

Internalized Fears

We read every day of the dastardly things men and women do to each other and to animals as well. Do we condemn all of mankind for this? Obviously not.

Similarly, we all know of the terrible things that have been done through the ages in the name of religion. Do we condemn all of religion for this? Obviously not.

Finally, we read every day of the negative effects technology has had on land, people, and animals. Shall we condemn all of technology for this? Apparently we should, at least according to some writers.

We, all of us in the developed countries, but especially those of us in the United States, are pictured as moths, hopelessly fascinated, fluttering helplessly about until we burn ourselves up in the light and heat of technology. Jacques Ellul, Herbert Marcuse, Louis Mumford, Norman O. Brown, Theodore Roszak, Kevin McDermott, and others are, in varying degrees, prominent proponents of this position.

The confusion, as I see it, is that the complainers and doomsayers are comparing life in industrialized countries today, not with conditions as they have actually been throughout history, but with some idealized society that has never existed, and may never exist. They are telling us, in essence, where we should go but not how to get there.

I have a feeling, too, that the fears expressed by some of our opinion leaders about the condition of our society and our future as a nation and a civilization are not objective ideas put forth by objective persons who have calmly evaluated all the evidence and then put forth their conclusion.

Rather, what we may be getting, at least in the more extreme cases, are the reflections of subjective, internalized fears—fears in the hearts of people who have not managed to come to grips with the world of today. I believe that at least some of those who are afraid of tomorrow are actually reflecting their own inability to cope with modern-day society.

I have no evidence or proof to support this rather strong statement. But it is surely obvious that one's personal feelings can strongly affect one's work. This was clearly seen in the case of George Orwell, who wrote one of the most depressing of all fictional antiutopias. He is reputed to have said that the novel, *1984*, "wouldn't have been so gloomy if I had not been so ill." He wrote a large part of the book while alone on a rain-drenched Scottish island, and finished it in a hospital. A little more than a year later, he died.[5]

Sometimes, too, people are afraid, and are ashamed of being afraid. A good way to handle this is to get others to "join the club." Thus the more people there are who have such fears, the less conspicuous the original fearers' fears are.

Shortly after World War II, Dr. Lester I. Coleman, an ear, nose, and throat surgeon, was called outside his office to help an injured youngster. What he saw was a pale, frightened boy lying on the ground, surrounded by six or eight adults. They made way for the doctor. He knelt down, gave some words of reassurance, and asked the boy how he felt. The boy whispered, "Fine—I'm okay." Dr. Coleman asked in astonishment, "Why are you lying there?"

"I want to go home, but every time I get up, they push me down again. They won't let me get up."

In Dr. Coleman's words:

The adults who had surrounded him had transmitted to each other the fear that the injured person (he had apparently fallen) must lie quiet until a magic word is spoken by someone who knows. The boy, by that time, was scared to death, though it had nothing to do with his injury.[6]

Point: We must distinguish between realistic fears and distorted, exaggerated, or unrealistic ones.

To help us in that admittedly difficult task, let's take a closer look at science and technology.

What Is Technology?

Everyone, in this country at least, recognizes that technology is a major force in modern society. Most, however, have the idea that technology refers only to large machines and complex electronic equipment, and that technology began with the Industrial Revolution some two centuries ago.

Some even feel that the United States and technology are somehow synonymous. But technology existed long before the United States was even a gleam in the eyes of the founding fathers.

Indeed, use of tools, a basic aspect of technology, is *not* a human invention. A number of animals, and even birds, use tools. An otter will often pick up a flat stone along with the shellfish or sea urchin it captures on the sea bottom. Lying flat on its back, it places the stone on its stomach and smashes the shell on it so it can get at the meat. A species of finch uses a pointed stick to poke insects out of the bark of trees. A wasp uses a large grain of sand as a hammer to makes its sand nest secure.

Even the idea that man makes his tools while animals simply use what is available is not true. The chimpanzee has been seen to select small twigs with great care, then pluck leaves and extraneous matter off it, and carry it around until finding a termite or ant nest, into which he pokes the twig.

And two British paleoanthropologists, Raymond Dart and Lewis L. B. Leakey, have found worked stone tools along with the remains of the very early ancestors of man, ancestors so ancient that their brains were no bigger than those of the apes of today.

Being afraid of technology is a little like being afraid of your right arm. For technology is simply man's way of doing things easier or faster, or of doing things (like flying) that he couldn't do otherwise. It is also man's way of extending himself, which he seems to do quite naturally. Thus the microscope and telescope are extensions of his eyes; the bicycle and car are extensions of his feet; the knife and spear are extensions of his teeth and nails; the hammer is an extension of his fist; and calculators and computers are extensions of his brain. Technology may also be used to provide aids, such as eyeglasses or hearing aids, to imperfect or failing senses. The British historian Arnold J. Toynbee maintains that "Our ancestors became human in the act of inventing tools." [7] And Herbert J. Muller, author of *The Children of Frankenstein*, suggests that "we are indebted to technology for virtually everything we own, including our minds." [8]

The control and use of fire may have been one of man's first uses of technology—for warmth, protection from other animals, and cooking. The case of the first two illustrate one important aspect of man. He is a generalist and is, in a sense, unfinished. He has no fur to keep him warm, and he cannot run from harm as fast as a gazelle; he is not feared because of his size, as an elephant is; he is not as strong as a bear; nor does he have the claws of a lion. With technology, however, he has been able to make quite good imitations and even improvements on these.

The third use, cooking, may be a good illustration of something else entirely, namely that of curiosity and a desire to improve his way of life.

It should be clear by now that technology is not just hardware. Thus, as the sociologist Daniel Bell points out, "the organization of a hospital or an international trade system is a *social* technology, as the automobile or a numerically controlled tool is a *machine* technology. An *intellectual* technology is the substitution of algorithms (problem-solving rules) for intuitive judgments." [9] According to Lewis Mumford, the organizations that built the giant pyramids and ziggurats of the ancient Middle East constituted the first "machine."

I do not wish to imply that there is nothing new in technology except for new devices and methods. There is, and it has to do with what might be called scientific technology. Putting it another way, we may say that technology as an art is old; technology as a science is new.

Science and Technology

Science is sometimes thought of as a dry collection of facts, a dusty collection of butterflies in a museum. But science is no more a collection of facts than a heap of bricks is a city.

The late Eugene Rabinowitch, who has been called "the conscience of America's scientists," [10] described science as "the supreme achievement of the cerebral cortex . . . a system of data and relationships covering vast areas of information derived from observation, analysis and manipulation of natural phenomena." [11]

That is as good a definition as I've seen, but it still suggests dryness. There is excitement in science, so we must look elsewhere. Maxim Gorky, the great Russian writer, once said that literature is the discovery of man; and being a process of discovery it is therefore a science.

Discovery. That's a key word. Science is also excitement, curiosity, a critical inquiry into the workings of nature, with man and indeed all the universe as its province.

Science has been called the search for the truth about nature.[12] But science does not, of course, hold exclusive rights to the search for truth. In their ways religion, philosophy, art, and literature are all seeking the truth. All depend upon observation, as science does, and even upon hypothesis—"this is what I think." The major distinction between these seekers and the scientist is the latter's contention that a hypothesis must be testable.

Science—the critical attempt to understand the workings of nature—has also been around for a long time. But it is only in the last hundred and fifty years or so that it has gone into partnership with technology.

One way of looking at this partnership is to see them as the halt leading the blind. Not only are both imperfect, but the interdependence has become very strong. This, indeed, has led to many people simply combining them into a hyphenated science-and-technology (which brings to mind the English king, William-and-Mary).

Clearly, however, simply lumping together these two diverse elements obfuscates more than it clarifies. Even the objectives are different: Science is investigating, measuring, understanding; it is the acquisition of knowledge.

Technology constructs, creates, accomplishes; it is the application of knowledge.

Ancient Greece, interestingly, was quite strong in science but not in technology, while the reverse was true of the Romans. What has come down to us from the Greeks, mainly, are systems of thought, such as Aristotelian logic, the Ptolemaic system of the universe, and many other lines of theoretical speculation,* while our legacy from the Romans has been more in the way of aqueducts, roads, buildings, and the dome.

I like to think of modern science and technology as the right and left feet of advancing man. Each step by one permits or even calls forth another step by the other. Joseph Black, the Scottish chemist (1728–99), is said to have come upon his theory of the latent heat of gases from observing the operation of whisky stills.† James Watt (1736–1819), an engineer who was a countryman of Black's, was aware of Black's theoretical work, and used it in developing his steam engine.

The invention of the high pressure steam engine around 1800 then led by various steps to the development of the laws of thermodynamics. These, in turn, were used by Rudolph Diesel (1858–1913) in his development of the diesel engine.

So there is clearly much interplay between science and technology. And modern technology can now be defined as the application of scientific knowledge to the solution of problems or the meeting of human needs.

But this meeting, it should be kept in mind, is relatively new.

---

* This may well be because the later Greeks tended to look down on what Plato and Aristotle considered "the base mechanic arts."
† It takes a great deal of energy to boil water. The energy content of a given amount of steam is therefore much higher than that contained in an equal weight of water.

The nineteenth century was a time of great ferment in the technological and industrial fields, but the developments were made largely by trial and error. The inventors and technical people in general were "tinkerers." Thomas A. Edison (1847–1931) may have been the last of the red-hot tinkerers. A genius whose list of inventions may never be equaled, he was not a booster of pure science; he was even heard to complain about his son, who seemed every so often to "go flying off into the clouds with that fellow Einstein."

As late as the First World War, science was still pretty much an outsider, even in the military world. James Bryant Conant, who was a chemist before becoming an educator, tells this story: Upon the entry of the United States into World War I, a representative of the American Chemical Society called on the then Secretary of War to offer the services of the society in the country's defense. The Secretary of War said the offer would be taken under advisement. A day later the representative was thanked for his kind offer but was told it was unnecessary because the War Department already had a chemist working in its employ!

Today, of course, the Defense Department does a great dead of technological development on its own, and is in addition one of the largest individual financial supporters of scientific research in university, institute, and industrial laboratories.

As a matter of fact, one of the major objections to science today is that it has become institutionalized. Yet there have been scientific institutions since the seventeenth century, financing experiments and disseminating results. But today much of science has become so big—both in total amount of research and in the size of equipment involved—that the scientist working alone in his private lab is pretty much a thing of the past.

Scientific research is now done by and large by large institutions that can supply the financial resources, the space and the equipment (scientific, technical, computer, etc.) that are required. By the same token, much of the work that is done today—radio and optical astronomy, high-energy physics, genetic research, and so on—could not even be considered without the sensing, measuring, electronic, and computing technologies that have become available.

Although many of us recoil at the very idea, science-as-an-institution is not necessarily bad. It permits advanced scientific results that could not be obtained otherwise— in medicine, sociology, psychology, archaeology, and other "human-oriented" disciplines as well as in weapons. And it permits something else: the end of science as a rich man's plaything.

Scientists used to be drawn from the ranks of the upper classes. It was they who had the needed resources and time, they who had the needed education for such play. The fact that they didn't have to "produce" meant that they could indulge in what we would today call basic, or fundamental, research.

With institutionalization, and with the fact that money and resources are being supplied from elsewhere, both science and technology have become less of an elitist occupation. People from all classes now supply the brains.

Today our "brains" are one of our most important resources. Our increase in real income over the past century has depended much more on our brains than on our brawn (conquest), or even on our natural resources (as with the oil-producing countries). Our per capita income over the past century has increased at an average of about 2 percent a year. Various studies indicate that the contribution of technological advance to increased productivity varies generally from two-fifths to five-sixths.[13] Certainly *that* is not something to be afraid of.

There are still human beings around who can say no to the dangerous and degrading aspects of modern life and yes to the rewarding ones.

The brilliant young writer Gene Youngblood, an arts-oriented person if there ever was one, put it well in his book, *Expanded Cinema:* "Technology," he wrote, "is the only thing that keeps man human." [14]

5 Social Progress and Technology

All progress is based upon a universal innate desire
on the part of every organism to live beyond its
income.
                    —*Samuel Butler (1835–1902), Note Books*

Progress, man's distinctive mark alone,
Not God's, and not the beasts': God is, they are,
Man partly is and wholly hopes to be.
                    —*Robert Browning (1812–89), Paracelsus*

If we can get over the fear of technology, and achieve
a better understanding of what it is and can do, we might
discover that it offers not only material benefits but the
chance for social justice. After all, the major push for
such justice in Western Europe came after, or at least
along with, the beginnings of modern science and the Industrial Revolution. There is no guarantee, but it seems
likely that the same will hold for the present-day less-
developed countries.

It is fairly clear that expecting social justice to filter
down from the top is a mistake. Progress along these lines
comes when people demand it. People in power, with

advantages over others, rarely relinquish privileges voluntarily.

It is likely that the only way this kind of change can take place is through the introduction of modern technology,* which in turn can raise the standard of living and increase the educational possibilities of the populace. Only then will the populace be able to think well enough of themselves to want to demand social progress; only then will they be capable of doing something about it.

The Nobel Prize winner Miguel Angel Asturias maintains that the Latin American's traditional "contempt for science and technology and the exaggerated value set on the humanities and scholastic learning have dragged us down. . . ." [1] He calls for the education of more engineers and fewer lawyers, more technicians and fewer doctors. And Mexican President Luis Escheveria Alvarez, in full agreement, observes, "Technological progress is today the best ally of the American Revolution." [2]

The process is beginning in South and Central America. Brazil, Mexico, and a few other Latin American countries are in the throes of the industrialization process.

Clearly, it is not necessary for a country to be as rich as the United States for its citizens to live decent, comfortable lives. But the question, everywhere in the world, seems just as clearly to be, How can the less-developed countries become developed countries? We castigate ourselves for producing what other countries are obviously very desirous of having. (At the very least, we have a large middle class as a result.) [3]

If the richer countries were to share everything with the poorer ones, the poor people would gain very little, while

---

* As we shall see, increased technology does not *necessarily* correlate with increased energy and materials use.

the rich ones would lose a great deal. The problem, of course, is that there are still many more poor than rich in the world.

So the only thing it makes sense for us to share is our knowledge, mainly of technology and administrative techniques, along with some seed money, tools, and so on.

What we are talking about, then, is interaction—interaction between the haves and the have-nots, the two classes into which the world seems to be divided.

The Haves and Have-Nots

Though the gap between rich and poor people within a society is as old as civilization, the great gap existing today between rich and poor countries can be said to have had its beginnings in the Industrial Revolution.

The Have and Have-Not Countries

The ground had already been prepared by the nautical explorations of such countries as Spain, Holland, Portugal, England, and France in the fifteenth and sixteenth centuries. These combined true curiosity with avariciousness, and led to, on the one hand, the first real knowledge of what the world really consisted of and, on the other, a cascade of riches. Precious metals assumed ever greater importance as a trade factor, giving a boost to the use of a money economy instead of the old barter system. The modern institutions of finance and credit also became fixed features.

By the eighteenth century a series of technological developments had been introduced, centering mainly around the application of power to jobs that used to be done by hand or animal. Later, interchangeable parts and the application of knowledge in new ways became important factors. All these, plus the new riches, led to both a demand for, and capacity for, increased productivity, both in industry and agriculture.

The result amounted to looting—though it was done a bit more subtly than in earlier days. The more advanced, more powerful West European countries, needing raw materials to feed their furnaces and machines as well as their bellies, simply turned the less-developed countries into plantations —hence the colonial empires that grew up in these centuries. The inhabitants who were willing to work were generally paid pitifully poor wages; the unwilling were simply dragged into slavery.

In an economic sense alone, the new abilities of the developed countries provided too much competition and often ruined native industries by underselling them.

A second reason for the great and growing gap lies in the very nature of the industrialization process. On the personal level, once the common people no longer supply all their own needs, as in hunting-gathering societies, or just their own food needs, as in agricultural societies, their dependence on manufactured goods rises rapidly. And if the country cannot supply enough of its own, then it must buy from outside, which often increases the price.

Over the years, aid of various sorts from the developed countries has had one particularly unfortunate effect on the less-developed ones. Our public health measures have helped lower the death rate while the birth rate has remained about the same. As a result, the growth rate has leaped—in some countries reaching 4 percent a year! India, with a growth rate of 2.5 percent, has some 14 million more mouths to feed each year—to say nothing of housing them, educating them, and giving them jobs.

Thus, even if a country is increasing its productive capacity, it may find that it is only holding its own, or even falling behind, in terms of the living conditions of its inhabitants. Unsatisfactory growing conditions have caused widening pockets of starvation in the last year or two. Yet many countries have simply run out of arable land. Therefore, the only hope for increasing food and fiber production is through improved technology. If no machinery were available to United States farmers, the muscle power of *all* Americans would be just equal to the energy needed to produce what we now produce with the aid of machines.[4]

If a group—nation, town, or tribe—decides that it is happier as it is, without modern technology, then we have no right to interfere. But there are no cases I know of where this has happened. In most less-developed countries, smoke from a chimney is a sign of progress, not pollution, and cars are eagerly sought, not abhorred.

The obvious advantages of technology are desired, even if the sometimes equally obvious negative aspects are not. Unfortunately, industrialization seems *not* to be an à la carte menu, from which you can pick and choose. It is a complete dinner. You may have some choice of dishes, but you've got to take—and pay for—the whole thing.

In a few cases, strong pulls have developed *within* groups between those who want to move ahead and those who, for whatever reasons, do not. This often happens when a group first becomes exposed to modern ways.

This is an example of fears that are perfectly realistic. Village elders, for example, stand to lose quite a lot if, due to the introduction of new methods, their knowledge of traditional methods of agriculture becomes obsolete. The workman, too, realizes that a new technology may make his job obsolete.

Culture Shock

The results of intercultural contact have often been surprising—sometimes successful, often disastrous. Sometimes the problem is simply that of difficulty in communicating with the less advanced group.* This is well illustrated in a story told of a U.S. Navy health officer on a Pacific Island during World War II. He was trying to show the health problems associated with the large fly population, and to interest the natives in an eradication program.

To make his various points, he used a large, foot-long model of a fly. He thought he was doing very well until one of his listeners said, "It's easy to see why you Americans are so concerned with flies. Yours are so big. Ours are just little tiny ones."

In a Haitian village, laborers were supplied with standard American shovels for digging a drainage ditch. The aid officer stood dumfounded as these otherwise quite dextrous people showed unbelievable clumsiness in trying to do the work. They made completely ineffectual attempts at digging—hitting each other and themselves in the back and sides. The aid officer couldn't believe his eyes. It was like the clown act in a circus.

The aid officer finally recalled that Haitians customarily work stooped over, and their tools are therefore invariably short handled and held near the blade. Cutting off two feet of handle did the trick.

* When I use the terms *less advanced* or *unsophisticated*, I am not implying that these peoples are inferior, stupid, or even simple-minded. They certainly are not. Indeed, their knowledge of plants, animals, and, in meat-eating groups, even anatomy, undoubtedly far surpasses that of most Americans. But in terms of modern technology they are surely less advanced than we are, and their thought processes are also, sometimes as a result, different.

Looking at the villager's response to the introduction of technology into his life can give us some insight into what bothers technophobic Americans. Anthropologist G. M. Foster says:

> For generations the villager has been able to show initiative only in the most limited areas. Small wonder that he often has trouble in making up his mind about something new. Moreover, not only does the villager (i.e., peasant) have little or no control over the basic decisions made from the outside, but *usually he doesn't even know how or why they are made.* The orders, the levys, the restrictions, the taxes that are imposed from the outside have for him the same quality of chance and capriciousness as do the visitations of the supernatural world. And the peasant feels much of the same toward both: he can plead, implore, propitiate, and hope for a miracle, but in neither case can he expect by his own action to have any effective control.[5]

We are reminded of Franz Kafka's haunting works, such as *The Trial* and *The Castle,* which indicate a feeling of helplessness and powerlessness, of being trapped by unknown forces in the process of trying to unravel the mysteries of life. But Kafka was a city dweller. Clearly, then, these feelings are common to both farm and city people.

The intrusion of the white man and his culture into the Pacific islands called Melanesia before and during World War II resulted in a strange kind of reaction called "cargo cults." [6] These arose independently but similarly in a number of islands. One common theme was that the world was about to come to a cataclysmic end; afterwards, earth would become a paradise. All the unpleasant aspects of life—old age, death, illness, and evil—would be gone. And, most significant for us, the material wealth of the white man—his "cargo"—would come to the Melanesians!

The story began to circulate that it was not the white man who was making all these goods but dead ancestors. The only thing they ever saw the white man do to obtain these goods was to make some signs on a piece of paper. Their priests and shamans often performed similar rituals.

The feeling began to arise that the arriving cargoes were meant for them but were being intercepted by the whites. In one case the islanders went as a group with a petition demanding that an arriving shipment be given to them— with results about as successful as you would expect.

The culmination of the cargo cult in each case was a kind of religious outbreak, almost a mania. Here is P. M. Worsley's description:

> The islanders throw away their money, break their most sacred taboos, abandon their gardens and destroy their precious livestock; they indulge in sexual license or, alternatively, rigidly separate men from women in huge communal establishments. Sometimes they spend days sitting gazing at the horizon for a glimpse of the long-awaited ship or airplane; sometimes they dance, pray and sing in mass congregations, becoming possessed and "speaking with tongues."
>
> Observers have not hesitated to use such words as "madness," "mania," and "irrationality" to characterize the cults. But the cults reflect quite rational attempts to make sense out of a social order that appears senseless and chaotic. Given the ignorance of the Melanesians about the wider European society, its economic organization and its highly developed technology, their reactions form a consistent and understandable pattern. They wrap up all their yearn-

ing and hope in an amalgam that combines the best counsel they can find in Christianity and their native belief. If the world is soon to end, gardening or fishing is unnecessary; everything will be provided . . .

Of course the cargo never comes. The cults nevertheless live on. . . . New breakaway groups organize around "purer" faith and ritual. . . .

"At this point it should be observed that cults of this general kind are not peculiar to Melanesia. Men who feel themselves oppressed and deceived have always been ready to pour their hopes and fears, their aspirations and frustrations, into dreams of a millennium to come or of a golden age to return.[7]

I wonder if this "ignorance . . . about the wider . . . society, its economic organization and its highly developed technology" does not reflect a similar ignorance on the part of many people in the United States. And while their reactions may not be as extreme, their fears of technology may nevertheless arise in a similar fashion. Recall the old saw about the city child who thinks milk comes from bottles and eggs from boxes: the average city dweller is usually just as ignorant of what technology really is and does as he is of the workings of nature.

This may be even more true for the educated person than for the laborer. For the blue-collar worker usually has at least some kind of intimate contact with machines.

Technology Transfer

Obviously the reason for giving aid to less-developed countries—or not giving it—is an important factor. Those who call for a "hands off" attitude, for instance, may well be reflecting a desire to keep the cultures in those countries as "museums," interesting places to visit when things get dull at home.

The negative aspects of the introduction of new technology can be physical as well as cultural.[8] In village India it has been common to cook indoors over an open fire using cow dung as a fuel. In the poorer homes, there is no chimney and there are few windows. The result, often, is a choking smoke, leading to many respiratory and eye ailments. The "smokeless chula," an inexpensive pottery stove and chimney, was introduced, intended to relieve these very real problems. Unfortunately, the roofs of the huts are thatched with dried grasses and leaves, which constitute a delightful food for a type of wood-boring white ant that, without the smoke, quickly proliferates and destroys the roofs.

Simple aid programs, therefore, often introduce either worse problems or additional problems. Does this mean that help should not be given? Hardly. But aid should depend on what the problems are, what other problems may or can arise: is the cure worse than the disease? Is the eradication of malaria worth the introduction of DDT? If we lived where malarial chills and fever debilitate large proportions of the population, we might feel differently about the chemical.

"Ceylon," writes author and consultant George T. L. Land, "is a case in point. In 1963 with less than 150 reported cases and no deaths from malaria, DDT spraying was discontinued. Five years later malaria had exploded to 2,800,-000 cases and claimed more than 12,000 lives, and DDT use was resumed."[9]

Yet the very poor farmer who, it would seem, would have the most to gain and least to lose, often turns out to be unwilling to accept a new technology. "Fear," wrote Professor Foster, "renders the poorest people incapable of trying new things . . . [they] cannot afford the risk of the unknown." [10] The best way, he says, is to find a few of the more enterprising persons, who almost invariably exist anywhere, and convince them to try the new method. Their success may be the touchstone of the whole program.

And then again it may not; for visible evidence of good fortune in a small village is often considered "proof" of guilt. For centuries the only way to increase production was to put more land under cultivation. "Clearly" there must be something wrong if the farmer increases his yield significantly on the same plot of land.

And just as clearly there is some of this kind of resistance in all of us too.

There are, as you might have concluded, no simple answers. One would think that self-determination is the answer: let the group in question make up its own mind. But sometimes self-determination, especially when mixed with technophobia, can produce unfortunate results.

I saw a good, and current, example of this on Monhegan Island, a beautiful, rugged, rocky island some twelve or thirteen miles off the shore of Maine. It is a small island, with perhaps fifty permanent residents; the main industries are tourism and lobstering.

For the tourists, Monhegan offers a glimpse of what life might have been like a century ago. For the island is, officially, devoid of electricity; one uses kerosene lanterns for light in most of the tourist inns. It is a pleasant and nostalgic experience. There are no cars on the island either, just a few trucks used by the hotel people for carrying luggage to and from the dock and for other chores. The area is hilly and the trucks are very useful.

I said there was no electricity *officially*, by which I mean that electricity has not been brought out from the mainland, nor has a central generator been brought in. This is because some of the residents want to maintain the old-fashioned feeling of the place. The innkeepers feel, with some justification, that if there is electricity officially then the lack of lights would be deceitful or dishonest.

On the other hand, they want the convenience of electricity for their refrigerators and other appliances. And even more urgent are the feelings of the lobstermen, one of whom told me, "You bet I'm for electricity. Life is hard enough here as it is. We need the electricity just to pump water."

He told me that before the arrival of electricity, there had been fears that the island would not survive as a community. "Ten years ago," he said, "we were down to one student in the school, and about thirty people on the island altogether. The young ones almost always left as they grew up. Now they are beginning to stay. That's the most important thing. And we've got fifteen or sixteen kids in the school now."

The no-electricity people have been able to keep out Electricity, but not electricity: individual generators have been brought in and lie scattered here and there. Thus, in this generally effective refuge from our noisy civilization, you can walk along the rocky paths in glorious silence, with nothing but the sound of the gulls in the distance, enjoying the serenity, when suddenly you will be blasted out of your reverie by the start-up of a generator not twenty feet from you.

This is self-determination or, as some might call it, paralytic stubbornness. The result in any case has been a poor compromise. With a more realistic attitude, a single large generator could have been brought in, buried somewhere in a remote part of the island, and used by those who wished to have electricity. With such a generator it would at least have been possible to maintain the truly lovely peace and quiet of the island, if not the darkness.

The Simple Life

Virtually all of those who extol the simple life are willing, nay eager, to partake of the fruits of modern civilization, not the least of which is the excitement of learning. This confusion of Arcadian and Faustian feelings was well illustrated in a college student I interviewed during my stay on Monhegan Island. This young man, whom we shall call Jack, worked there during the summer to help pay his way through school. He had the best of both worlds. He was enjoying his summers at Monhegan, but was also enjoying college and was eagerly looking forward to a career in biology. He thought he might even go on for an advanced degree.

Jack really loved the island; he loved the slower pace and the way of life, and talked seriously about settling down there after he got his degree or degrees.

I asked him how he reconciled the idea of advanced education with that of the simple life. The entire island, I pointed out, subsists on lobstering and tourism, with a few artists snuggled in here and there. Would he really be able to put his knowledge to use if he did settle on Monhegan later on?

Jack was a highly sensitive, intelligent, and thoughtful human being; yet he admitted that this conflict had never occurred to him. He lapsed into silence, and thought for a while; then smiled weakly and asked if we could go on to something else.

It would seem then that these summers were a welcome respite, a change, from his regular activities. But could he take a full-time life of lobstering or catering to the whims of sometimes unpleasant customers? And where would he find the wherewithal to practice his profession on his own?

The Need for Myths

The Yir Yiront, a tribe of Australian aborigines, had not, at last report, "graduated" to the boat stage. To cross a body of water they still hold onto a log and swim.

Yet there are dangerous crocodiles, sharks and other fish, sting rays, and Portuguese men-of-war in the surrounding waters. This limits their range of fishing, on which they depend heavily for food. They know of boats, for only about forty-five miles away there is another tribe that has mastered the art of making bark canoes, and the Yir Yiront have had contact with them.

Why, then, did they not take on this admittedly useful technology, especially since the necessary materials exist in their own environment? They do not have canoes, says Prof. Lauriston Sharp,

> because their own mythical ancestors did not have them. They assume that the canoe was part of the ancestral universe of the northern tribes. For them, then, the adoption of the canoe would not be simply a matter of learning a number of new behavioral skills for its manufacture and use. The adoption would require a much more difficult procedure: the acceptance by the entire society of a myth, either locally developed or borrowed, to explain the presence of the canoe . . ."[11]

In other words, for these people at least, new developments have to fit into a myth system. How much similarity is there between such peoples and ourselves? How significant are myths in our own society?

Has our technological progress outstripped not our social but our literary and maybe even artistic capabilities?

Are we really a rational society?

6 The Turn to Irrationalism

Irrationally held truths may be more harmful than reasoned errors.

—Thomas Henry Huxley, The Coming of Age
of the Origin of Species, p. xii

The traditional African Zulu will not look into a dark pool for fear that a lurking beast will grab his reflection, and that he will thereby be deprived of his soul.

A quaint superstition, we chuckle. Yet how many of us still say "God bless you" when someone sneezes? This ritual derives from a belief of the ancients that the soul was blown out with the sneeze, and that some magic term or activity was needed to get it back. (In some languages the words for *breath*, *soul*, and *spirit* are closely allied or are even the same.) We could say that this is no longer superstition but mere habit or custom. Yet ask yourself if you feel just a tiny bit nervous if you sneeze and no one says the magic words.

How many of us still shiver when a black cat crosses our paths, especially in the evening? And how many will not walk under a ladder—even though there is a better chance that the ladder will topple over and hit us when we are walking around it than that it will simply collapse on us when we are walking under it.

It is said that insurance salesmen prefer to do business at night, not because the "man of the house" is usually there at that time, but because it is dark then and most people are more fearful at that time!

And just try to find a building where the thirteenth floor is called the thirteenth floor.*

Every bookstore has its occult section. Indeed books, articles, movies, and lectures about spiritualism, occultism, and so on, can almost always count on a good sale.

There are an estimated 12,000 fulltime "professional" astrologers and 175,000 part timers in the United States.[1] A recent Gallup poll indicates that 32 million Americans believe in astrology.[2] Virtually every newspaper in the country has an astrology column. In Great Britain more than two-thirds of the adult population read their horoscopes.[3] Many people will not make any important move without consulting "their" astrologer first.

One hundred and fifty thousand Americans list spiritualism as their religion, and millions more believe one of the major tenets of spiritualism,[4] that we can communicate with the dead.

William J. Peterson, from whose book, *Those Curious New Cults*, many of these figures are derived, adds that there are ten thousand self-proclaimed witches in Germany. He also reports a witch's claim that there are 10 million practicing witches in the United States, which he calls "witchful thinking." He reports estimates of between 10,000 and 100,000 Satanists in this country alone,[5] and

* A highly inaccurate survey of *new* skyscrapers along Third Avenue in Manhattan revealed that in half of them the thirteenth floor was called fourteen!

that another cult, Scientology, has several hundred thousand believers in the United States and 2 or 3 million worldwide.[6]

In addition to this group, there are the many Eastern religions and various other cults that have taken hold in the country, particularly among the young.

A major objective of many of these cults is to get across the idea that rationality, science, and technology—and civilization itself, at least as we know it—are a package of trouble, that the real way to happiness lies in eschewing all of that. Use intuition rather than rationality or logic to make decisions; eliminate individualism; become one with some larger something in the universe; and, most important, get back to, become one with nature. This last aspect, of course, means getting rid of or cutting down on technology, for technology obviously is not natural.

Under pressures from those urging strict interpretation of the Bible, publishers have been downplaying Darwin's theory of evolution in their biology textbooks, especially at high school level and below. The *New York Times* reports: "A study of biology texts by Dr. Judith V. Grabiner and Peter D. Miller found that the biology texts commonly used before the Scopes trial included more on evolution than is included in many books in use today." [7] Dr. Thomas H. Jukes of the University of California maintains: "Lots of kids are coming out of school totally ignorant of one of the most important concepts in science." [8]

We can gain some insight into the overall situation by considering an idea put forth several decades ago by Talcott Parsons, who suggested that irrationality is to thought systems what entropy is to physical systems. In physical systems there is a natural tendency toward disorganization—iron goes to rust, never the reverse—and entropy is a measure of that disorganization. Similarly, suggested Parsons, the natural tendency is toward irrationalism; it takes work to move toward or to maintain rationalism. Freud referred to the "frail crust of reason."

Rationalism and Irrationalism

Let us examine some of the typical ideas of the irrationalist, antitechnology movement.* A major tenet is that nature is good, beautiful, and so on, while man's works are ugly, bad, and so on. Yet what is natural? People have been getting arthritis for a long time; it's quite natural, while the artificial hip joints some of the badly afflicted have gotten are, by this definition, unnatural.

Similarly, many doctors in the nineteenth and early twentieth centuries didn't want to give anesthetics to women giving birth because women were *supposed* to feel pain—it was natural.

Another idea of the irrationalist is that we should let the heart rule (because the rational approach has made such a mess of things). Ruling from the heart sounds like a good idea. But while I have seen no studies on the subject, it seems very likely that most of the conquerers who killed, raped, looted, plundered, and burned their way through history were ruling from the heart.

It has even been suggested that nazism was an outgrowth of an antitechnological and antirational youth movement that took place in Germany in the 1920s and 1930s.[9] Both fascism and nazism glorified nature and downgraded the idea of intellectual or rational thinking. But nature in these ideologies meant the "survival of the

---

* I can perhaps be challenged for attaching these two together. Certainly not all antitechnologists are irrationalists, and vice versa. But the combination is more likely to be true than not, for reasons that will, I hope, become clear as we proceed.

fittest." It's not hard to guess who they felt were the fittest. Hitler was an occultist; and nazism was to him a religion that was going to supplant Christianity. As Pascal once said, "Men never do evil so completely and cheerfully as when they do it from religious conviction." [10]

A third aspect of irrationalism has to do with depth of meaning. Science and technology are somehow thought of as shallow, lacking not only humanness but depth. Charles Frankel, professor of philosophy and public affairs at Columbia University, writes of the myth that "persists among many educated people that rational inquiry thins out the world or deprives human experience of its extra dimensions of meaning." [11]

A friend of mine maintains that scientists and engineers can't appreciate the beauty of anything because they are always too busy picking it apart.

But is one's appreciation of music necessarily diminished if he knows that the lengths of strings that produce a note, its fifth, and its octave are in the simple ratio of 6:4:3? This was, after all, known to the Greeks.

I asked Philip Morrison, professor of physics at MIT, and book editor for *Scientific American*, about this. He thought about it for a moment, then quoted from a poem by Keats:

> Do not all charms fly
> At the mere touch of cold philosophy?
> There was an awful rainbow once in heaven:
> We know her woof, her texture; she is given
> In the dull catalogue of common things.
> Philosophy will clip an angel's wings.

"But," exclaimed Professor Morrison, "we know quite a bit about the rainbow, and I still think it's wonderful. I think it's more wonderful because we do understand it now. We can see that perfect circle and the set of the colors. That's there; no one can take it away from us. But we look for the second rainbow. And sometimes we see it, you know, under funny conditions. We think about, we cultivate, that rainbow. Some people even take an infrared picture of the rainbow, which nobody ever saw; it's been there all the time and nobody ever saw it.

"That adds a much greater dimension to the purely visual picture that the artist sees. The problem comes when you tell people that the rainbow is no longer interesting and no longer beautiful because now we know how it works. That's nonsense! That's when you're in bad, bad shape. And that's what bad science education has done.

"I think it's a mistake to identify science with rationality completely, because it has growth, it has novelty, it has puzzles. The whole interest in it lies there. If it doesn't have mystery, it's no good."

Indeed, biologists are fond of quoting a "law" that says something like, "Under the best laboratory conditions, and with carefully planned and executed experiments, you may be sure that the experimental animals will do as they damned please."

Although most people think of science with awe, they miss most of its magic and mystery if they are ignorant of its workings—which is just the reverse of mysticism, spiritualism, and other such activities. In magic and alchemy there is usually some single, simple gimmick —the philosopher's stone, the fountain of youth, the *I Ching*, the astrologer's charts, or whatever. To scientists, the workings of nature are rational and comprehensible, but still require continuous work and searching.

Thus it is more likely to be the irrationalists, the non-scientists, the mystics, who claim to have all the answers or, as they might say, the important ones. As one of my readers once put it in a letter to me: "I will give you the address [of a religious "science" organization] because I believe that you are sincere in your search for the truth. A scientific background shouldn't be a barrier to logical and inspired thinking." I should hope not!

The idea that rationality somehow detracts from all things is extremely pervasive. In a certain sense, it is true. A monstrous bee or spider is a logical impossibility. The creature would break its wings or legs at the first flight or step. Why? Because as the length of the insect goes up its weight increases as the cube of the dimensional increase, while the cross section of the legs only increases as the square. Thus if the length increases one hundred times the weight goes up a million times, while the cross section of the legs only increases ten thousand times, which means that each unit of supporting material must suddenly be able to support one hundred times the weight it used to.

Is it better not to know this? Since it makes about half the "science fiction" movies utterly ridiculous, does it remove one more aspect of "fun" in our lives? But is life really more interesting because a certain segment of the motion picture industry produces the same movie over and over again, merely substituting a giant bee for a giant spider this time, and a giant ant the next? Is there so little creativity in art that our lives are impoverished by no longer being able to "enjoy" such fare?

Irrationalism in the Past

The American tradition has *not* been hospitable to science, or even to intellectualism. Aside from a few general periods—such as the turn of the twentieth century and the post-World War II era—there has been little real support for science itself. The scientist in America has long and generally been regarded with the same mixture of awe, respect, and fear as has the shaman of old. He was, and is, an intellectual, for whom the common people have traditionally harbored a mixture of fear, respect, and dislike. Intellectuals have always been different. Thus classical musicians were "longhairs" (when long hair was not in fashion), and scientists were "eggheads."

The antiscience and anti-intellectual feelings of the past are at least explainable, if not understandable, for this was a country of pioneers who needed, more than anything else, strength, courage, and practicality. New ways of doing things, new machines, and cheaper goods due to factory production (technology!) were all welcomed. Science in the United States has by and large only been welcomed as an adjunct or stimulus to technology. A widespread appreciation of the intrinsic beauty of science, an aesthetic awareness, were no more in evidence a century ago than they are now.[12]

Nor is the idea that science and technology, taken separately or together, are leading us down the road to perdition a new one. Indeed, there are throughout history examples of a kind of pendulum swing between rationalism and irrationalism, and between optimism and pessimism, particularly among the artistic and literate peoples. The scientific and intellectual achievements of the seventeenth century, for example—the thinking and writing of people like Newton, Descartes, Spinoza, Francis Bacon, and John Locke—coalesced into a belief in universal order and the comprehensibility of the universe. This became the Age of the Enlightenment (self-described, interestingly). Eighteenth-century Europe particularly began to see a lively questioning of authority—scientific, political, and religious—and an emphasis on empirical method: depending on experiment and experience, rather than on theory and belief, for gaining knowledge.

Then there came in the late eighteenth and nineteenth centuries what seems to be the inevitable reaction. This time it was called *romanticism*, an astonishing period of excessive sentimentality, of a revolt against rationalism, of complaints about where we were heading. Romanticism has been called the secular equivalent of earlier mystical and occult rites.

In our country the transcendentalists of the mid-1800s tried to fit science into the religious mold they believed was the measure of the world. And with this group of irrationalists, we come to the real meaning of the words *rationalism* and *irrationalism*. Rationalism in the modern mind is generally associated with reason; the implication then is that irrationalism is an absence of reason. But the transcendentalists never doubted the quality of their reason. According to George Ripley, reason was "the faculty of directly seeing the divinity of Christianity." And the Englishman John Locke, though not a member of the group, maintained that one could proceed from a set of intuitively known propositions, using deductive, logical steps, and produce proof that God exists.

Individual intuition was—and to some people still is— thought to be the highest form of knowledge. One dream may be worth one hundred experiments. Knowledge comes to us in three basic ways. Some we are born with, some comes to us in an unknown fashion, and some we obtain in the usual scientific way. This last way, sometimes called the *empirical approach*, was considered to be the least important of the three.

Toward the end of the century William James, the influential American philosopher, said there are three ways of establishing belief: reason, experience, and feeling. But for Ralph Waldo Emerson (1803–1882), one of the best known of the transcendentalists, reason was the door to ethical and religious truth.

The transcendentalists felt that the common man was better equipped to find the road to Truth than the veriest intellectual, that rural Reason was far superior to any other kind of rational or scientific reasoning. Yet, in the irony of all ironies, their philosophic writings tended to be so complicated, so tortuous, that no common man could read them and thus learn what a paragon he really was!

Irrationalism Today

Charles Reich tells us that "the young drug user of today just plain 'knows it.'" He explains, "It might take a Consciousness II person [e.g., organization man] twenty years of reading radical literature to 'know' that law is a tool of oppression; the young drug user just plain 'knows' it." [13]

This sounds very much like the knowledge of the heart, the power of intuition, or whatever. In any case, we have heard it before.

The counterculture, Reich also tells us, is "deeply suspicious of logic, rationality, analysis." [14] Who wouldn't be suspicious of something he knows nothing about?

Theodore Roszak, a professor of history, welcomes the counterculture's flight from reason, and even calls it "the saving vision our endangered civilization requires." [15]

Roszak attacks the rationalists for ignoring the dream world.[16] But they do not ignore it; it is just that their approach is different. They prefer to do the analysis while they are awake.

Interestingly, new developments in science and technology are often latched onto as evidence of paranormal phenomena, or phenomena beyond the realm of science, and also to pry money out of gullible people. At various times in history, magnetism, electricity, x-rays, and radioactivity were all used in this way. Think of the electroencephalograph (EEG), first publicized in 1929. Electric waves being generated by the brain! Irrationalists had a field day with that one. "Obviously" it was the explanation for thought transference. Today the machine is used routinely in diagnosing epilepsy, brain tumors, and damage to or degeneration of the tissues of the brain.

Currently a whole new group of phenomena, and some not so new, are playing the same part. Kirlian photography, biofeedback, changes in the normal functioning of the body, trance states, and other phenomena are offered as proof of the existence of paranormal phenomena. In Kirlian photography, for example, high-frequency discharges of electricity from the surfaces of objects are photographed. The reason irrationalists are interested is twofold; one, when living things are photographed (fingertips, leaves, etc.) the emanations are unpredictable; thus the element of the mysterious is present.

Second, some interesting occurrences have taken place. In one case, Samyon Davidovich Kirlian, a Russian electrician and discoverer of the effect, could not get a good result during a demonstration; the photographs showed only dark spots and clouds. Suddenly he began to feel ill, showing symptoms of a chronic circulation disorder.

His wife took over the demonstration and was able to produce good pictures; the equipment was obviously in good working order. They then alternated; again pictures of his hand "showed a confused, chaotic pattern of energy, blurred and cloudy. Valentina's [his wife's] hands showed a clear pattern of the discharging stream of energy, the colored flares bright and sharp." [17]

Could such pictures show evidence of illness, *before* illness strikes? Is this evidence of something beyond science or is it merely a kind of sensitive technique for picking up cues that have been there right along but not seen up to now? It seems reasonable to assume the latter and, further, that in time the technique will enter the medical chest.

Much harder to account for is a different phenomenon. A Kirlian photograph of a leaf that had had an edge cut off showed up normal except for one slight surprise: the photograph showed the original outline of the leaf in toto! [18]

The fact that we in the West are now doing some of the things that Eastern mystics have been doing right along (slowing the heartbeat, lowering blood pressure, and so on), is offered as proof that these phenomena are beyond the pale of science—that is, because they have been done first by mystics.

But that's the whole point of science, to explain the unexplainable. The scientist is interested in understanding the phenomenon, the technologist is putting it to work. And *we* should be interested in seeing that, when it is put to work, man will somehow benefit from it.

Why the Cults?

Why do so many bright, intelligent, sensitive, honest people turn to cultism, mysticism, and other forms of irrationalism and anti-intellectualism, as well as antitechnology?

J. Eric Holmes, of the University of Southern California medical school, suggests that "such beliefs are part of normal mental development. Perhaps one learns to be critical of flying saucers and telepathic plants only by going through such intellectual fads and being personally disillusioned." [19]

But the answer to Why? has many more aspects.

First, and I think most important: Life, as some wise sage once pointed out, is not a bowl of cherries. It is, for most people at least, a period of toil and difficult decisions, punctuated by disappointments and pain, and merely garnished with moments of pleasure here and there.

The rationalist says, That's the way it is; face it and make the best of it. Death, the total cessation of life and consciousness, is inevitable. You can perhaps affect your own destiny, but nature is rational and wends its way regardless of how you try to shunt it onto another track.

The spiritualist, on the other hand, says the dead are not gone; a lost loved one can still be contacted. The astrologer says you can affect the course of destiny by doing or not doing certain things on the basis of whether the time is propitious. How the astrologer knows better than you whether the time is propitious remains a mystery.

Yet even the traditionally religious person, not the cultist or mystic, is in reality doing something very similar. A large percentage of prayer consists of asking for some favor or other. There is surely a great deal of comfort in the idea that you can sway the course of destiny by somehow convincing whoever is at the controls to intercede for you.

There are also the admitted emotional and psychological needs of people, what Joseph Sittler of the Divinity School at the University of Chicago calls "the mind's longing to know wholeness and depth." [20]

But science, technology, and rationalism do not, cannot, offer this. They offer, rather, small steps. They stake out a small claim and then proceed to work it. Irrationalism offers the whole world in a basket. It is, admittedly, a hard offer to turn down.

René Dubos feels that mysticism provides satisfaction and even enjoyment. And among very young people, we do indeed see them "playing" with Tarot cards and Ouija boards, and "believing" in Dracula and "spirits."

Among the older youth, turning toward mysticism and irrationalism could be an outgrowth of these earlier experiences. It may in addition result from feelings of helplessness; it may seem easier to influence and control cards or even stars than college administrations and the Pentagon—once you latch on to the right method.

Thus it is not surprising that such activities seem to thrive in times of stress, in times of anxiety. This happened several times in Europe, and particularly in the 1600s, after an epidemic of plagues. Hasidism, a mystical Jewish movement, arose among the Jews of eastern Poland in the eighteenth century after a long period of religious persecution. It happened in Germany after its collapse in World War I; and it seems to be happening again in the Western world, where threats from nuclear war, pollution, mass starvation, and the population explosion are much in the news.

Is this a time of stress? One way to test the general feeling of the times is to look at the literature, both fiction and nonfiction. In general it can surely be said that today's literature is pessimistic, even apocalyptic.

By way of comparison, let us travel back in time some two millennia and take a look at the apocalyptic literature of the Bible. Both the Old and the New Testament have plenty of it.

One objective of apocalyptic literature—prophesying a cataclysm from which only the righteous would emerge—

was to encourage the Jews of 200 to 100 B.C. to "hold on" until the promised destruction of their persecutors (e.g., Book of Daniel in the Old Testament). Things would improve—in the next life if not this one.

The early Christians had similar problems with the Romans and, during the period of about A.D. 50 to A.D. 350, produced their own brand of apocalyptic literature, of which the Revelation of Saint John the Divine is a good example.

Today we are also seeing a widespread production of apocalyptic literature.[21]

But there is a difference. What is missing today is the return, the Second Coming, the triumphant emergence of the "good guys."

Thus, at exactly the time when we should be our coolest and most rational, we are told that rationalism is the culprit and should be exorcized. At a time when man has developed enormous powers, he is at the same time seen as hopeless and helpless. At a time when, in material terms at least, there is a realistic hope that all the plagues of mankind could be routed by scientific technology, the New Apocalyptics give up hope.

Is there not something wrong here?

As Arthur Schlesinger has remarked, "Reason without passion is sterile, but passion without reason is hysterical."[22] But the question then arises: "How shall we strike a balance? Shall we have one part of the people research, develop, and manufacture, while the rest wine, wench, and spiritualize?"

Must we have a division between the followers of Apollo and those of Dionysius?

Surely the truly fulfilled person is one who manages to combine these two very different approaches to life. The ancient Greeks emphasized man's need for both Love

(Eros) and Reason (Logos). Surely there is room in our lives for both.

It is true that all of us have emotional needs, ignored at our peril. But that they can only be satisfied by mystery or superstition or religion is a vast oversimplification. For some the sound of a running brook may do far more to soothe irritated nerves than the loveliest hymn or the words of a guru. For others a sunset, the arranging of flowers, a tea ceremony, may do the trick. Skiing or hiking in the woods, painting, fiddling, or developing pictures may provide the link with one's feelings that is sometimes lost in the quest for material things.

It is likely that we are *all* more or less irrational at times. A problem may be that Western man, and particularly the rationalist intellectuals, have no institutionalized method of giving vent to their irrationalism.

Art in its broad sense—fiction, drama, dance, and poetry—can give us a good picture of where and what we are, and what is missing in life; it can define what is good and spur us toward the infinite, wherever that may be. It speaks to us on a "gut" level.

But our artists may have come to believe—perhaps in line with the New Apocalypticism—that they have lost the power to affect our daily lives, and so have taken refuge by turning away from life as the hermit seeks refuge from the difficulties of interaction with other people.

Today's artist, no less than any other member of our society, may be a victim of rising expectations, to my mind the one really hopeless disease introduced by technology.

7 Technofears—And What About Them

The world has no longer any mystery for us.
—*Marcellin Berthelot (1885)*

One of the commonest fears of technophobes is that technology is an unstoppable juggernaut. The basic idea is that technological developments have a kind of life of their own and will, when once they take hold, burgeon and grow like a cancer, killing off the good life around them. They grow, it is believed, regardless of the feelings and desires of the people, regardless of how wasteful they are of money and resources.

The will of the people and the good of society, it is suggested, have no bearing on the matter at all—the individual citizen is a helpless pawn in a giant chess game being played by "them."

We all can quote instances of this. During World War II, for example, someone got the idea of building flying Dewar flasks, large aircraft fitted out to carry liquefied gases at cryogenic (extremely cold) temperatures. Several, as I understand it, were actually built, at an extremely high cost.

Granted this was foolishness and a waste of money,

though it might be considered a kind of technological WPA. If there is no real need, however, such things fall by the wayside eventually. Those flying vacuum jugs were never put into operation, and have gone the way of the dodo bird.

But when there is a need or use for something, it slides into wide use with no trouble at all. Toward the end of the nineteenth century, John B. Dunlop was seeking some way to provide a smoother ride for his son's tricycle. He got the brilliant idea of wrapping an inflated rubber tube around a wheel and came up with the pneumatic tire, which he patented in 1888. The tire was an instant success: there was a need for it.

Nathan Rosenberg, in his interesting book, *Technology and American Economic Growth*, points out that "it was rural households which were in the forefront of the adoption of the automobile for private family use in the early days of the automobile industry. Urban families were, by and large, equipped with reasonable public transportation alternatives." [1]

Mechanization often arises because of some shortage or other, mainly one of labor. Rosenberg also tells us:

> The long-term rise in agricultural productivity was dramatically accelerated in the years after World War II. A strong inducement to mechanization was imparted by the growing demand for labor in the non-farm sectors during the war, which raised wages and led to a large-scale movement of labor out of agriculture." [2]

Things like snowmobiles, dune buggies, minibikes, fancy cars, and so on are made because they are bought. And they are bought because they satisfy some kind of need, want or desire. If they do more harm than good —and that can be a difficult determination to make—it is on that basis that they must be banned, and not because someone who has no use for them says they are useless.

It is true that the new development may blossom or metamorphose into something entirely different, as happened with Dunlop's pneumatic tire. Without it the automobile industry might still be an infant and air pollution would be far less of a problem than it is today. One wonders what would have happened if someone looking down from on high had warned Dunlop, "Better not go ahead with your idea. You wouldn't believe what it's going to lead to."

To many people, technology is just "too big for its britches." In reality, however, technology is only big enough to satisfy the wants and needs of government, industry—and the public!

All of this comes down to more than an academic point. Techonology, whatever it is, is big, and even growing. But that does not mean it will not respond to direction.

The actual decisions for its use, of course, are made by politicians, by businessmen, sometimes by the electorate, rarely by technical people. A scientist suggested that an atomic bomb could be made; but no scientist or even group of scientists could have done it. Equations do not explode. It took the multibillion-dollar Manhattan Project to do it. It took the naturalist-politician Adolf Hitler to create the proper climate for it. A group of scientists who knew what was going on tried all in their power to prevent the bomb from being built, and another tried to prevent it from being dropped. One of the major points made against J. Robert Oppenheimer when his loyalty to the United States was being questioned in the early 1950s was that he had opposed the hydrogen bomb project in 1949—although he had headed up the lab that built its predecessor, the fission bomb.

Oppenheimer obviously had a fear of technology—but it was a rational fear of a fearful weapon, especially one being proposed at a time when we were not at war.

There is also an unfortunate tendency to confuse technology, which is technique and ability (and will-less machines), with something that does appear to be a virtually unstoppable juggernaut, namely the "defense" establishment.* In 1961 President Eisenhower warned: "Until the latest of our world conflicts the United States had no armaments industry. . . . This conjunction of an immense military establishment and a large arms industry is new in American experience."[3] Even worse, adds Seymour Melman, the old military-industrial complex that President Eisenhower warned about "did not have a built-in mechanism for enlarging its own decision power; the state management [now centered in the Defense Department] has precisely such a mechanism."[4]

In other words, the military establishment is a bureaucratic establishment, and like all such establishments it has a will, and a desire for self-preservation, if not aggrandizement. It may be so powerful that it can't be budged † but that's not because of its technology. ‡

Let us therefore try to keep the excesses of the military separate from those of technology. The latter do exist,

---

* The confusion is well illustrated by the fact that, on the one hand, we are told "the system is out of control and is going to take all of us with it," while on the other that it is tightly controlled by the military-industrial complex, big business, or whatever. How can both be true at the same time?

† Even the military doesn't always get its way, of course. One idea that was *not* put into practice was to loft a mirror satellite that would reflect sunlight down onto Vietnamese territory during the night to prevent their carrying out activities under cover of darkness.

‡ As Arthur C. Clarke once put it, "If there is ever a war between man and machines, it is easy to guess who will start it." Though large systems do tend to keep going through sheer inertia, they can be, and have been, scrapped as their need or value decreases.

but they constitute a phenomenon of an entirely different order.

Specifically, many large-scale decisions (e.g., to build the A-bomb and nuclear submarines) are more military than technological, just as the decision to put a man on the moon was political, not technological.

A true decision can only be made by a person or a group. And in spite of all the talk about the creation of wants by the public relations and advertising industries, it nevertheless remains true that the automobile was seen as the answer to a large number of transportation problems and deficiencies that existed at the turn of this century.

That we do not need two-ton monsters to take us three blocks for a loaf of bread is surely true. But that cars filled a need is shown by the basic fact that we still—three quarters of a century after their invention, and in an age of "explosive change"—have not found a more convenient way to get people from place to place. Ford and General Motors did not create the automobile industry; it created them. Only later did symbiotic grandiosity set in.

Dehumanization

The idea that science and technology are dehumanizing is a common one. We read, for example, that the early and mid-sixties was "the time of the historic robot-like Cape Kennedy countdowns." [5]

But this is like calling the excavation of a foundation by dynamiting a robotlike operation. It is necessary to clear the area when the time comes for the explosion. But prior to that, and in the background, there is plenty of humanity involved. And anyone familiar with the space flight countdown knows that this is no robotized operation.

Thousands of employees are involved in every phase of the procedure.

Part of this reaction has to do with the fact that many who fear science and technology are convinced that research is a cold, calculated, computerlike affair carried out in a perfectly logical, step-by-step fashion; that it is, basically, humorless, unaesthetic, and pitiless. Overall, the development of science and technology is seen as a kind of grand march down the boulevard of logic. Everything is perfect.

Actually, the true breakthroughs are made in exactly the *reverse* manner. Indeed, it almost *has* to be this way, for the logical move, being merely another step along a fairly well-established road, could be made by almost anyone. The breakthrough is seeing a road where none was even faintly discernible before. Thus there is a touch of the marvelous, the mysterious—compounded, of course, with large doses of hard work, heartache, failure, and tenacity. (Thomas A. Edison is said to have tried more than six thousand different materials in his search for an electric light filament.)

In 1885, one could hear the lament of a famous chemist, Marcellin Berthelot, that "The world has no longer any mystery for us."

But then, in rapid succession, elementary particles, radioactivity, and x-rays were discovered, knocking that complacent attitude right out the lab window. Within a few more decades new discoveries by Planck, Einstein, Heisenberg and others had replaced absoluteness with relativity, certitude with probability.[6] The kind of science the scientific illiterate is thinking of—complete, arrogant, knowing—has been out of date for the whole of the twentieth century.

For example, the uncertainty principle of Werner

Heisenberg says you cannot know accurately the position and velocity of a particle at the same instant. Thus the idea presented a century or so earlier by the French mathematician Laplace—that if we could only know the positions and velocities of all particles at a certain instant, we could know exactly what would happen at all times in the future—was seen, finally, to be not only impractical but theoretically unsound.

Thus those who complain that scientists know, or think they know, everything are showing their ignorance of what science really is. What makes science interesting is not what is known, but what is *not* known.

As for the idea that technology is dehumanizing us, I wonder what the word *human* really entails. Is heavy physical labor humanizing? Is bending over a rice field ten hours a day humanizing? Is using slaves to do your work humanizing? Is shoveling coal or chopping wood all day long humanizing?

There is, of course, a strong school of thought, exemplified by such people as Thoreau, Gandhi and A. D. Gordon, who maintain that manual labor *is* humanizing.[7] I would counter with this fragment of an old story I heard somewhere. A poor man comes to a rich man and, obviously starved, pleads for work. The rich man takes pity on him and gives him the job of cleaning out his privy.

A visitor arrives and is appalled that the rich man should degrade the poor man like that. He argues that by assigning such work he degrades both the poor man and himself, while if he did it himself he would find it ennobling. "There are some things, sir, a man should do for himself," he exclaims.

But why should *anyone* have to do that kind of work when a machine can do it, or technology can make it unnecessary?

The things that make our lives more than a struggle for subsistence—art, music, literature, theater, and dance, for example—are provided by people who have been released from that struggle thanks to technology. Those who do-it-themselves have similarly been released by technology—in simpler societies as well as our own.

Those who *want* to perform physical labor can easily find opportunity to do so.

And how about the women in our "dehumanized" society? Stephanie Harrington says that *Ms.* and *Cosmopolitan* magazines have risen on a "tide of 'revolution' —cultural revolutions set in motion by a technology that has reduced the amount of time women have to spend on household chores and by a rising living standard that has sent more women out to work to supplement family incomes." [8] How about these women, who have been freed from the drudgery of long hours of clothes washing, food preparation, hand sewing, and so on? Do they feel they have been dehumanized?

Again, when complaints are made, it is likely to be the better-off women who make them; they are the ones who, in earlier days, had servants to do all this for them.

Some interesting thoughts on the relationship of women and technology are contained in a paper by Joan A. Rothschild of the University of Lowell. She points out that

> much feminist analysis and experimentation is concerned with alternative ways of organizing sexual relationships, reproduction and childrearing. The literature ranges from the more radical proposals of a [Shulamith] Firestone to debiologize women by freeing them completely, through technology, from childbearing—including artificially creating human beings

and making it possible for men to bear children—to science fiction writer Ursula LeGuin's future society of neuter/hermaphrodites who may experience motherhood *and* fatherhood in their lives." [9]

A different aspect of the woman/technology question is seen in the fact that women are still typically less interested in science, mathematics, and engineering than men are. A lady photographer I know complains that our cultural bias has, as a result, made it impossible for her to discuss the technical aspects of her craft with her compeers. She is working hard now to remedy her lack of science education. Her "natural" bent in this direction was apparently subjugated to the idea that the important thing was to be a housewife and do the "natural" things that women do. She obviously does not feel that science and technology are inherently dehumanizing.

Medicine Kills

An interesting word is sometimes used in medicine; it is *iatrogenic*, which means "doctor-induced," or "doctor-aggravated." We all know of cases in which a treatment has caused another, perhaps even more serious, problem than the one it was intended to cure. Suppose a person given penicillin for an infection turns out to be allergic to this drug and becomes gravely ill or even dies. Is penicillin thus a killer? Should it be banned?

Considering that it has saved probably thousands or even millions of times the number of people it has harmed, we can see that getting rid of it, tearing it out of the arsenal of medical techniques, would be foolish indeed. But that it should be used less freely is undoubtedly true.

The use of penicillin and other antibiotics can be thought of as an example of the overall use of technology: careless use can be dangerous. With antibiotics there is not only the danger of an allergic reaction to a drug, but the factor of immunity is also involved. If a drug is used often enough to treat a person for nonserious ailments, it may no longer have the desired effect when needed for a serious illness.

Yet large numbers of people insist on taking antibiotics for virus infections such as the common cold, against which antibiotics have no effect.

Bacteria and other microorganisms also seem to build up immunity to these drugs; penicillin has had to be continuously modified ever since its introduction during World War II to keep up with the organisms it is intended to control. These organisms reproduce so fast that their chances for mutation are great; if a single mutant in a group of virulent bacteria happens to have immunity to a drug, it can all by itself produce a whole new generation of immune organisms. Thus the medical research operation must be one of constant work and attention.

The problem is that people are more afraid of machines than of medicines, though they might well be more wary of the latter. In *Consumer Reports* we learn that

> more than 575 different tablets, liquids, powders, lozenges, gums and pills compete to soothe our stomach complaints. [This is a] $200 million market. . . . One specialist reported that in a single hospital he had witnessed five cases of gastro-intestinal hemorrhage caused by taking the product (Alka-Seltzer) for stomach problems. In each instance, Alka-Seltzer had been used by the patient to treat

symptoms arising from ulcers or other serious dis-
orders of the stomach. That temporary "cure" for
pain had only aggravated the damage.[10]

One could blame the drugs, and many people do. Much
more sensible, it would seem, would be a way of controlling their use. But balancing the benefits and hazards
of medicines and drugs is a difficult task at best. Here is
what Alvan R. Feinstein, M.D., professor of medicine and
epidemiology at the Yale School of Medicine, has to say
on the subject:

> Consider the reports of the thrombophlebitic * haz-
> ards associated with the use of oral contraceptive
> pills. Some workers doubt that the hazard even ex-
> ists, but, assuming that it does, what statistical evi-
> dence has been promulgated for the other side? Are
> there any statistics about the benefits of avoiding
> pregnancy for women whose lives would have been
> endangered or substantially worsened by another
> child? Are there any statistics on the human gratifi-
> cation achieved during sexual intercourse without
> mechanical contrivances and without the fear of
> pregnancy? Are there any statistics about the number
> of marriages saved or redeemed by the sexual libera-
> tion that 'the pill' provided for the marital partners?
> Data about human joys and comforts are not easy
> to get, so we ignore them. We engage in an un-
> balanced science that quantifies risk but not bene-
> fit.[11]

The word iatrogenic, defined earlier as "doctor-induced"
or "doctor-aggravated," may also refer to hospital-induced

---

* Thrombophlebitis is a condition characterized by inflammation of a
vein wall combined with a blood clot in that area. (H.H.)

problems, which again we have all heard something about, if not experienced. Books and movies like *Hospital, Such Good Friends, Interne,* and so on make for good, if frightening, entertainment, showing, as they do, the horrendous practices carried out in some hospitals. But it must be kept in mind that they are good entertainment because they are different from the norm. If what they depict were the norm, dog bites man, it would be of less interest than man bites dog. In the vast majority of cases, moreover, it is people—doctors, nurses, orderlies—who cause trouble because of carelessness, greed, or fatigue. Remember that the country doctor also make mistakes.

Admittedly, muckraking has its place, especially if it creates a desire among us to do something about a problem. But many stories about hospitals and doctors are likely to develop exaggerated fears in us. So it is important that we at least try to determine what is true and what is exaggeration.

There is more to the hospital-fearing syndrome. Hospitals are larger and better equipped today than ever before. But "better equipped" implies the use of more complex and larger machines. Hence the hospital has become threatening from its greater complexity as well as its greater size.

The question of size brings us an interesting point. In factories, large size may well be an advantage, and may lead to economies. But this may not be true of establishments that deal with people, such as schools and hospitals. Schools that are too big may be overwhelming to the student; similarly, some patients can stand a large hospital and others can't, and that is important because the patient's attitude can affect his health and recuperative powers.

My own reaction to large hospitals is a positive one. I tend to feel safer there. Many of my friends who have to be hospitalized also prefer to go into one of the large New York hospitals, such as Columbia Presbyterian, one of the largest in the world. Because of its great size, it is equipped to handle almost any emergency. And the human body, being the complex mechanism it is, can come up with the most unexpected emergencies. As someone once put it, just being alive is a risk.

But in a large hospital with all that equipment available, there is often the temptation on the part of the doctor to use it—to give additional tests "just to be sure." For some of us that's all right; but for the fear-prone it may cause more damage than the tests are worth. The trend toward increasing numbers of malpractice suits and larger awards may also be causing many doctors to practice "defensive medicine."

Thus far it is still the doctor who calls for the use of these medicines and tests, and who should be using some of his humanness to make proper decisions in such cases. If he or she cannot, or does not, then maybe it would be better to have machines do it. Medical people (like any technical group) must also be careful to use nonfrightening language, to try not to show off, when talking to the layman.

On the other hand, people ignorant of the ways and capacities of medical science often force medical people into untenable positions. Patients want answers. It is the patient with whom the physician spends the most time to whom he is least likely to be able to say, "Your problem is X and we are going to cure you by doing Y."

But how many times have you heard someone snort: "Doctors! They don't know what they're doing. First they try this, then they try that. . . ." The physician would like to be able to say, "I'm not sure, but I'd like to try . . ." The patient, not willing to accept the fact that medical science is not omniscient (and certainly not infallible), objects to being used as a guinea pig. So the physician may very well decide that it's better to say, "It looks like *R*, and these *T*s should help." And considering the interplay between mind and body, it may even be better for the patient to think that the treatment will help.

In the process of gathering material for this book, I was struck often with the remarkable, indeed incredible, things that were being done, particularly in the field of medicine. I read stories that made me wonder about the oft-stated suggestion that nature is soft and beautiful, and man hard and ugly. There was the case of a young girl whose body became tied up in knots, and who suffered excruciating pains in the process. For several years she was "accused" by psychiatrists of, in essence, doing this to herself to gain attention.

The case was finally brought to the attention of a brain surgeon who had been doing work with cryogenic (very low temperature) probes. The operation he had developed had its dangers, as does any operation, most particularly one on the brain, and the surgeon hesitated to work on this preteen-aged girl. But the mother said to him, "Either you make the effort and try to relieve our daughter by this brain operation, or we're going home to turn on the gas. We mean it—the whole family. We can't stand it any longer." [12]

The child was a victim of one of the most tormenting diseases known, dystonia musculorum deformans. An operation was done, and considerable relief was obtained.

Would it have been better if Cailletet and Pictet, and Dewar and Kamerlingh-Onnes had not previously experimented with the behavior of materials at low temperatures? If technical people did not move on from there, "fooling around" with various techniques of producing low temperatures? The beginnings lie back in the 1850s, when Joule and Thomson showed that if a gas is allowed to expand freely its temperature drops slightly. Who would have dreamed that the trail would lead to use in exquisitely delicate brain operations?

What is beautiful about this particular technique is that the suspected problem areas can be probed and frozen a bit at a time to see if that is indeed the section of the brain that is causing trouble. If it is, then that tiny bit is destroyed.

Yet suppose someone had been able to decide at an early stage that low temperature work was useless and pointless, that that money would have been better spent on something useful, like finding a better way to clean manure off the city streets.

It is my feeling that the major difference between human society in the developed countries and that of the social insect—bees, termites, and ants—is that the individual, every individual, is important. And clearly medicine has played a major role here. Families no longer have to have many babies because a significant percentage were almost sure to die before reaching adulthood.

So each human has become precious. Is that bad? Inhumane? Technology has made it harder and harder for us to turn away when someone is sick and in pain. In the "good old days," the doctor or the parent would mumble "It is God's will," and that was that.

Slippery Slope

With all that has been achieved in the field of medicine, there still remains for too many people a firm belief in "the slippery slope of science." The implication is that if we allow work to proceed on in vitro fertilization * (objective: to help women who cannot conceive naturally), then we will surely end up with bottle babies à la *Brave New World;* if we permit amniocentesis † to continue (objective: to detect seriously defective children early in pregnancy), we will surely end up with a eugenic program in which all "undesirables" are prevented from being born.

Amniocentesis has been performed several thousand times, with the result that a considerable number of mongoloid fetuses have been detected and aborted, thus perhaps saving the parents a long life of misery. Yet no one has shown any desire to include in this program children with hangnails or other such undesirable characteristics! Amitai Etzioni, professor of sociology at Columbia University, points out: "The record shows that practicing professionals *and* citizens at large can redraw the line, and at a rather sensible point . . . most doctors and laymen favor amniocentesis for detection of severe illness and adamantly reject it for sex choice.

"Most important," Prof. Etzioni adds,

> the facile humanist disregards values other than the taboos he is so anxious to preserve, values that would be violated if we were immobilized by fear of innovation. Humaneness cannot be guaranteed by putting a stop to all scientific work in an anxiety-provoking area, but by carefully assessing the multiple applications of scientific discoveries—promoting some, discouraging others, and foregoing still others. We cannot be spared the choice.[13]

* Fertilization of an egg outside the uterus, then reimplanting it.
† Use of a needle to obtan a sample of the amniotic fluid, which contains cells of the fetus, and analysis of the fluid for certain diseases and defects.

Invasion of Privacy

Yet another common fear is that technology will bring surveillance and what Theodore Roszak calls "inner manipulation" to the point of perfection, that we will all then be totally at the mercy of the "state"—whoever or whatever that may be.

But who is going to do this? Will we have half the world spying on the other half? Taking turns, perhaps? Will a chosen few be able to keep a cover on everyone else? Are not the primitive methods used in the totalitarian countries today, namely, of getting people to keep the state informed of any suspicious activities of their neighbors, really much more sensible and just as effective?

Or maybe, as hinted, machines will be doing this. But on what basis? A few key words, perhaps, such as *anarchy, burn, communism?* Anyone saying one of them, and overheard by the hidden "bug" in every room, will be immediately hauled off to the clink?

In *1984*, George Orwell's masses are the downtrodden, the manipulated, the miserable. In the United States and other industrialized countries today, the vast middle class is accused of being smug, complacent. But if a society can make a large majority of its constituents *think* it is happy (without resorting to the happiness pill, soma, as in Huxley's *Brave New World*), then it is doing remarkably well. If the society can't or hasn't fooled the large majority in this way, then the suggestion of Roszak, Marcuse, and others that technology has enabled mass man to be manipulated just doesn't hold water.

There are numerous consumer, nonestablishment, and underground papers, magazines, and books being freely published and distributed. New technology in the form of copying and reproduction machines of all kinds has made that possible. Some two thousand small, independent book and magazine publishers constitute one of the liveliest cultural activities of our day.

Indeed, it would seem that new technology has provided more opportunity than ever for the dissemination of anti-establishment ideas. Some people make their statements on photographs or on film. The lightweight Porta-Pak television camera/recorder has brought a whole new world into being on tape.

For many people today, depersonalization and loss of privacy are both associated with the growing use of the computer. When an escaped killer is tracked down with the aid of a computerized data bank, we all applaud, but the next day we are complaining again about loss of our own privacy. It's the old "we-them" thing all over again; unfortunately *we* and *them* are too often interchangeable. And if the law enforcement agencies get carried away, well, we have to find ways to prevent this from happening.

We *should* worry. There is indeed the danger of unnecessary files being kept, of people who have no business knowing certain things gaining access to such files; of more and more of them being kept unnecessarily and compared because it has become easier to do so with computers and photocopy machines. These are real dangers.

But society has always been a race, or maybe a tug of war, between civilization and chaos, between order and freedom. What is needed are laws, good laws, the best we can create. Sometimes new developments, like data banks, make us more cognizant of these dangers, and something is then done about them. There are laws that have recently been passed in some states, for instance, that require school administrations to permit parents to see their children's records. Thus damaging comments by teachers or administrators, which used to travel safely along with the child from grade to grade, are brought out into the open.

All of this requires a heads-up, wide-awake attack by legislators, with support and prodding from an alert, aroused, knowledgeable citizenry. An understanding that it is not the problems that have changed much, but rather the techniques for dealing with them, can help us overcome the fear that prevents action and understanding.

A bill being considered now in Congress, called the Comprehensive Right to Privacy Act, shows the difficulties in balancing an individual's right to privacy with the needs of organizations setting up the data banks. (A fascinating coincidence: the number of the act is H.R. 1984!) The act will, for example, require the organization to make available to any individual with a file contained therein, in a form comprehensible to him or her, all personal information concerning that person, the nature of the sources of the information, and the names of recipients of personal information about the data subject. Clearly this could create quite a load on the organization.

The experience of the Food and Drug Administration under the Freedom of Information Act of 1966 is suggestive. This act was recently amended to make more records and documents available to the general public and to expedite the handling of requests for information from federal agencies. As it turns out, however, there are few inquiries from these sectors. Most, and there are many, have come from corporations seeking information about competitors, and from lawyers hoping to gain information regarding liability suits.[14]

Clearly such a data bank would have great social utility.

But there remains that difficult question: Is it possible to set one up while still protecting privacy? And if not, how much of their privacy are people willing to barter for increased disease or crime prevention and control?

As for depersonalization, the middle class is now, thanks to computers, getting a taste of the treatment the lower classes have been getting for ages. And they don't like it. Maybe they'll do something about it. A caution, however. Computers are a way of increasing productivity. It may be that we have to give up some of the personalized service we have become used to—which used to be provided by lesser mortals than ourselves—in the interests of raising the sights and living status of all. This, after all, is the avowed aim of all humanists.

The Destruction of Civilization

I am not against criticism—of our government, of our society, of our media, or whatever. But there is a difference between destructive and constructive criticism, as every parent and every teacher should know, and as every person should be taught.

We *are* poisoning our environment; but we are also beginning to do something about it.

And even now when we are, we hope, at the worst of our pollution, the numbers of those who have become sick or who have died because of it is still minuscule, in spite of our far greater numbers, when compared with the numbers of those in preindustrial times who suffered or died from starvation, cold, drought, disease—and polluted water.

There are some writers who are concerned about man's desire to control nature. William Leiss argues that "men regard science as increasing their control over nature and consequently their happiness and well-being, thereby blinding themselves to the fact that by vastly magnifying their ability to indulge their collective passions they threaten the destruction of civilization." [16]

But a considerable amount of civilization consists precisely of those aspects of life that have been wrested from nature with the aid of technology. Where does survival end, and comfort begin? Where does comfort end, and luxury begin? Is there any doubt that yesterday's luxury is today's requirement?

Must we go back to the condition of the Pakistani villager who reels with flood, drought, locusts, disease, and every other manner of plague? That's nature, too. I much prefer my present condition, even if it means domination of nature in its negative sense. Besides, to decry man's domination over nature is to exaggerate. Hurricanes, hail, cyclones, tidal waves, floods, and other natural disasters still rampage at will. The Federal Disaster Assistance Administration tells us that the forty-six natural disasters that struck the United States in 1973 caused more than $1.2 billion damage in thirty-one states, and that seventy-five thousand stricken families required federal assistance. In 1972 things were even worse, with forty-eight major disasters causing $3.5 billion in property damage alone in twenty-three states.

Some of this, it is true, we bring on ourselves, as when we build on flood plains and otherwise disregard nature. What is needed is a more sensible and intimate relationship with nature. We shall see that this is not only possible; it is probably necessary. But it does not necessarily mean less technology. Rather, it means a different kind, and it presupposes a more intimate relationship between people and their technology as well.

Technology as a Weapon

Technology has put weapons of mass destruction into our hands. But neighboring groups have been at each other's throats since the beginnings of history, and though swords kill individuals while nuclear weapons kill hundreds of thousands, it makes no difference to the victim whether he is dispatched by sword or bomb. I myself prefer to hope that the sheer massive power of nuclear weapons will continue to make the potential users more careful, as it seems to up to now; thirty years have passed since nuclear weapons were used over Nagasaki and Hiroshima. Just as the Vietnam war sickened a whole generation, maybe a frightening enough specter of nuclear war will make those at the top less trigger happy.

I hope none of this will be taken as support for nuclear weapons. *All* weapons are to be feared and should be banned. Still, throughout history weapons technology has been an instrument for change. For example, the cannon made the old castle-fortress system obsolete, and led to the nation-state. Today the existence of super weapons and accurate delivery systems has made the nation-state as vulnerable as the castle became five hundred years ago.

Shall we react with prophecies of doom? Or can this new challenge lead us to reexamine the very idea of war? Can it lead to some new thoughts on how to use the new technology as a way to insure global peace?

The ideas of a Washington-based group called War Control Planners, Inc., suggest a route. Starting from the basis that unilateral disarmament could not be sold to the American people, and that it is not a good idea in any case, they propose instead the use of technology to act as a sort of world sentinel.

First step would be a so-called global information co-operative—a large-scale, open-to-the-public system, linked to military and civilian earth satellites and other global intelligence sources. The objective would be to maintain a public inventory of potential public danger, whether from threat of pollution, drought, hurricane, blight, or war. In other words, *any* threat to the general well-being of human beings everywhere would be detected and, hopefully, checked.

There are already earth satellites in orbit that can observe and record via computers the amount of rainfall, condition of soil, growth rate of crops, attacks of disease, and yield of crops. Conditions can also be recorded regarding forestry, fire, wildlife, mineral resources, weather. On command, the computer can be made to print out, for any country, what it has seen. It can therefore also report on troop movements, mobilization of equipment, the direction of movement of a military force.

No one would be forced to join in this cooperative venture, but the hope is that as countries see it working, they would join it voluntarily.[17]

All countries today disclaim aggressive intent and arm for "self-defense." To the extent that this is true, a potent spy-in-the-sky capability will cut down on arms activity. To the extent that it is not, aggressive intent will at least be clearly revealed.

A nuclear accident in weapons or reactors is of course also a possibility, as is the potential for madness in any nuclear area. Along this line John T. Edsall of the biological labs at Harvard suggests that

> nuclear fission plants will be enormously attractive objects for sabotage and blackmail. A well-placed charge of explosives, in the midst of one of these huge

concentrations of radioactive material, could blow into the air enough radioactivity to be carried by the winds over thousands of square miles, and perhaps render large areas uninhabitable for decades. The twisted minds, and the savage emotions, that could lead to such acts, seem utterly alien to most of us, but . . .[18]

But. But why are such acts so often thought of in terms of new technology? At a major football game, there may be a hundred thousand people in close proximity, surely easy prey for such a mind. Similarly, it would be relatively easy to poison the water supply of a large city and potentially affect millions.

Should we eliminate a central water supply because of this possibility? Should we stop attendance at football games?

The point is that modern technology has indeed enlarged the opportunities for madness, but it has by no means created them. There have been central water supplies since biblical days, and perhaps before; but there have been, to my knowledge, no such incidents (although a year or so ago someone in West Germany threatened just such action unless some large sum of money was paid to him).

It is true, of course, that terrorism of various kinds has been a terrible and tremendous problem—attacks at airports, the Olympics, and so on. But again, it hardly seems fair to pin the blame on technology. Even ignoring the justness or unjustness of the cause, it seems possible that the terrorists would be less likely to go ahead with their plans if they did not feel confident that they could always find refuge in some country or other.

A basic fear is that a terrorist group could steal waste plutonium (a by-product of American reactors), and use it to construct a crude, though still powerful, bomb. While it is true that they would need only about ten pounds of plutonium, and could fabricate a bomb if they were mechanically inclined and had access to a machine shop, the theft itself is more difficult than it seems.

The Federation of American Scientists explains: "The hijackers would have to steal at least six tons of cargo to have enough plutonium for a crude bomb in view of the weight of the containers; the fabricated fuel required to yield enough plutonium for the bomb would itself weigh 3,300 pounds. The gang stealing the material would be separating out the ten kilograms while a fantastic search proceeded for its machine shop (which would have to include the ability to handle the heavy containers) in which the activity was taking place." [19]

Due to recent publicity, security precautions for all such materials have been tightened considerably.

Another problem that must be faced is nuclear contamination of the environment. As a matter of fact this is a multiple problem; one aspect concerns potential accidents with consequent spread of radioactivity; another has to do with storage of wastes; and a third concerns transportation of fuels and waste, for example, from reactor to reprocessing plant and back, or from reactor to storage areas. One approach that has been suggested to overcome these and other problems as well is nuclear "parks"—large-scale, enclosed areas containing several reactors, plus facilities for reprocessing of the used fuel, and perhaps even capabilities for waste disposal. This would provide a comfortable distance between population and radioactivity, and would cut down on the transportation of these dangerous materials. Reactors cannot explode, so there is no additional danger in grouping them.

The storage problem has not been solved, but would not

seem to be insoluble. It has even been suggested that these wastes might one day be a valuable commodity.

Dr. Dixy Lee Ray, past chairperson of the Atomic Energy Commission, maintains that the complete bundle of wastes from the hundreds of nuclear plants that may be operating by the year 2000 can be stored by methods now in development in one single acre of land! That would be God's Little Acre indeed.[20]

The sensible approach to all of this, again, is an educated, unemotional call to lawfully constituted bodies that all necessary steps be taken to do what has to be done—including, if necessary, stopping the diffusion of technology that has been clearly demonstrated to be dangerous, until we can handle it more safely. This is surely preferable to making doomsday predictions of inevitable catastrophe, which only make all us turtles pull in our heads.

Another basic problem with many attacks on technology is that they are emotional. This is indicated by the fact that not only are science and technology accused of all kinds of wrongdoing, but the attacks are often personalized as well. Dr. Jerome B. Wiesner, now president of MIT, reports that "as Special Assistant for Science and Technology to President Kennedy, I was frequently attacked for the many alleged wrongs of the scientific community. In fact, many critics of American science and technology seemed to hold me personally responsible for the things they didn't like." [21]

The motto of a great scientist might be used to advantage by the rest of us. Madame Curie believed that "nothing in life is to be feared—it is only to be understood."

Results are achieved only by those who believe that

action is possible. The positive-thinking person moves forward, aware of the strengths of his society as well as its weaknesses, determined to improve things, or at least not to contribute to gloom, pessimism, and self-defeat.

8 *Time to Cut Down?*

At Magny's dinner. They said that Berthelot had predicted that in a hundred years of physical and chemical science man would learn to know the atom, and that with this knowledge he would be able, at his will, to dim, extinguish or relight the sun like a Carcel lamp. Claude Bernard, for his part, is said to have announced that with a hundred years more of physiological knowledge we would be able to make the organic law ourselves—to manufacture human life, in competition with the Creator.

For our part we did not raise any objection to all this talk, but we do believe that at that particular stage of scientific development, the good Lord, with a flowing beard, will arrive on Earth with his chain of keys and will say to humanity, just as they do at the Art Gallery at five o'clock, "Gentlemen, it's closing time."

—*Edmond de Goncourt and Jules de Goncourt:*
Journals, April 7, 1869

"Let us not," we hear, "use yet another technological 'fix' to get us out of the situation the last one got us into." But how well have *non*technological approaches done? We have as a good example the highly topical energy dilemma. The energy researcher tries to find new sources of energy, more efficient ways to use it, and cleaner ways to produce it; the alternative, or social, approach calls for cutting down on its use.

Granted, the energy picture would now be much brighter if we, and our antecedents, had modulated our energy orgy. But that hasn't happened, has it? Everyone blames everyone else—oil companies, politicians, automobile manufacturers, the Arabs—for the energy shortage. The only ones who come out clean, so the public likes to believe, are the public. But we, as well as our public officials, have been warned over and over again that we had to conserve. Forty years ago C. C. Furnas, an associate professor of chemical engineering at Yale University, wrote:

> Oil is an unknown quantity—1910 predictions said no oil by 1920—1920 said no oil in 1935, but in 1935 oil is so plentiful and cheap that they do not water stock any more—they oil it. It is unpredictable but by no means limitless. At our present rate we are undoubtedly going to begin having our oil troubles in a few decades at the most.[1]

Over and over again we have been told that our accelerating use of energy and resources is going to make trouble for us, that in the last half century man used as much material and energy as he did in all his years to date. The response has been a huge and continuing yawn—after as well as before the 1973 oil embargo. Use of gasoline seems to be inversely related to the size of lines at service stations. Even the fact that the price of gasoline is roughly double what it was before the oil embargo seems to have had little impact on use.

People still sit in shirtsleeves and, indeed, short sleeves, complaining about how cold it is in a library, store, or home. And often they will push the thermostat up above 70° F. instead of putting on a sweater.

Turning back the habits of 210 million people is no easy task. John C. Fisher, in his book *Energy Crises in Perspective,* maintains that as early as 1850 we were using more fuel per capita than Great Britain and West Germany are today! [2]

Then, too, in our governmental system, at least, politicians have to get elected, and supporting strong, unpopular measures can be political suicide. We could tax gasoline prohibitively or ration it, or we could tax big cars heavily. But it is almost two years after the embargo, and we are still waiting for action.

Part of the answer to our energy problem obviously lies in reeducating ourselves and retraining ourselves to do with less. Clearly we need to "think conservation"—and we must tell our legislators we will support them in that area. But we will surely also need the much-sneered-at technological fix.

My wife and I, for instance, have tried to do our little bit by keeping the thermostat at 68° during the winter. We wear heavy sweaters; the house is insulated and has storm doors, as well as double glazing (Thermopane) or storm sash on all windows; I have stopped all the openings around windows and doors; and I have even installed a humidifier. The upshot of it all is that we have been somewhat uncomfortable through two winters now. The alternatives are to sit in front of an open fire in the fireplace (where 80 percent of the heat goes up the chimney), or to raise the thermostat a couple of degrees.

Our experience along these lines—mild discomfort—leads me to wonder whether those who call for cutting down actually do so themselves. Perhaps they are like the driver who would like all cars to disappear—except his own.

It isn't that there is something particularly wrong with *us,* aside from our having been spoiled for a while. Keith Pavitt, of the University of Sussex in England, points out: "Raw materials were scarce in Europe so that European innovations tended to be resource-saving. . . . In the U.S., however, labor was relatively scarce, so that innovations tended to be labor-saving." [3]

But every society has used as much of its resources as it could, combined with as much technology as there was around, to make itself as comfortable as possible. The northern Europeans have been shivering through their winters for thousands of years—until, that is, the advent of central heating.

Nevertheless, with some five percent of the world's population, we *are* using a third to a half of all energy and materials. Though we are paying for them, and in some cases very dearly, clearly that cannot continue. There are several billion people "out there" who are hoping to move up the economic ladder—and who want their share of that energy and those materials. It would seem, then, that finding more efficient and less polluting methods of production is extremely important under any circumstances.

Restraint of Technology?

Does this mean there should be no holds on technology? The beginnings of an answer are given by Vice Admiral Hyman Rickover, career navy man and technologist of the first order, the man whose bulldog determination had much to do with the development of the nuclear submarine.

He points out, first, some differences between science

and technology: "Science has great authority . . . technology cannot claim the authority of science . . ."

He adds: "The methods of science require exclusion of the human factor. They were developed to serve the needs of scientists, whose sole interest is to comprehend the universe . . ."

And finally: "What scientists discover may shock or anger people—as did Darwin's theory of evolution. But even an unpleasant truth is worth having. . . . Science, being pure *thought*, harms no one; therefore it need not be humanistic. But technology is action, and often potentially dangerous action. Unless it is made to adapt itself to human interests, needs, values and principles, more harm will be done than good. Never before, in all his long life on earth, has man possessed such enormous power to injure himself, his human fellows and his society as has been put into his hands by modern technology." [4]

It is technology, then, not science, that should be regulated. But what does the admiral mean by *regulation?* He does not mean turning the clock back, or even stopping it. Trying to stop it is like telling a ten-year-old boy that you don't like the idea of his growing out of his pants, that his appetite is big enough as it is.[5]

It just doesn't make sense.

Suppose, nevertheless, that we really tried to get rid of technology. We would find that there are probably not enough natural materials around to take care of our needs, let alone our desires. There are just too many of us at this point. For one thing, 70 percent of textile fiber used in the United States is now man made. The Celanese Corporation maintains that if we were to attempt to revert to use of only natural fibers—mainly cotton and wool—we would require at least an additional 13 million acres of farmland,

and that this would be a poor trade-off in today's hungry world.

Do we include man-made fur in the ban? Shall we go back to widespread use of the fur of leopard, cheetah, and other such endangered species?

Natural lumber is a fine building material. But plywood is for many purposes stronger and more dimensionally stable; it uses our forest resources more economically and costs less. Particle board uses material that was long wasted. But plywood and particle board couldn't be made without the adhesives provided by chemistry.

It is probably not an exaggeration to say that human survival now requires much of the technology we engage in. Five hundred years ago the country was supporting about a million Indians in their way of life. They shared the land with vast herds of buffalo, and with various other wildlife. But that way of life requires about one square mile of land per person. The United States has some 3 million miles of land area. So if we were to do away with modern technology entirely, we would also have to do away with about 207 million Americans.

An agricultural society, of course, would not require such Draconian measures. But then farming—anything other than pure subsistence farming, which has its own problems—also involves technology.

One interesting objection to technology has to do with our machines breaking down, usually at the most inconvenient times. But:

Point number one. If the machine is of any use at all, what time would be convenient for its breakdown?

Point number two. Machines are not perfect. Nor are we. Anything that is used gets wear and tear. We probably do better than machines because we are self-repairing. We can make self-repairing machines, but they would be very costly. We already build in redundancy in places where this is really necessary, for example, in space flight.

Point number three. If you are foolish enough to buy an electric can opener without having a simple mechanical one on hand, that's your problem. If you build a new house and air condition the whole thing, ignoring the possibilities of sensible design, shade, orientation with respect to prevailing breezes, cross ventilation, and so on, then you deserve what you get.

Remember too that many of our useful machines are the front line of defense between us and the malevolent aspects of nature. For nature is not always benign.

Here is what Eric Hoffer has to say about the idea:

> I spent a good part of my life close to nature as migratory worker, lumberjack and placer miner. Mother Nature was breathing down my neck, so to speak, and I had the feeling that she did not want me around. I was bitten by every sort of insect, and scratched by burs, foxtails and thorns. My clothes were torn by buckbrush and tangled manzanita. Hard clods pushed against my ribs when I lay down to rest, and grime ate its way into every pore of my body. Everything around me was telling me all the time to roll up and be gone.[6]

Growing Pains

Until quite recently, a small town could pretty truthfully be described as a big city that never made it. There were few places indeed where chambers of commerce and booster clubs weren't looking for more population, more industry, more everything.

Now, with almost terrifying suddenness, the pendulum

has swung in the other direction. Thanks to the fear psychology so prevalent today, growth is out; no-growth is in.

We like to think that everything that happens today is unique to our time; but the recommendation for a no-growth society is an old and oft-repeated one. It is implicit in many utopias, as it almost has to be; for growth means change, and if change is allowable or desirable then the utopia obviously was not perfect. John Stewart Mill, the British philosopher and economist, wrote approvingly in the nineteenth century of the "stationary state."

The no-growth idea is the new utopian bandwagon, a dangerously simplified answer to a horrendously complex set of problems

There are two basic problems. One is a rapidly growing population: each person, especially in the developed countries, is a consuming, waste-producing unit; the more people there are, the more consumption of energy and natural resources, and the more pollution.

The second problem is an increasing worldwide consumption per capita. Lester Brown, author and president of the Worldwatch Institute, claims that worldwide growth in annual consumption of goods and services is about four percent a year, and is about equally divided between population growth and rising individual affluence.[7]

In other words, even if world population growth could be cut to zero, the world economy would still be growing at the rate of 2 percent a year.

The "obvious" answer, then, is to cut both population growth and economic growth. The clarion calls have come to be Zero Population Growth (ZPG) and Zero Economic Growth (ZEG). To these I would add, by implication, Zero Technology Growth (ZTG).

Even the most optimistic technophile will admit that

there has to be a halt in population increase. With about 4 billion people in the world, that growth rate of 2 percent means an additional 80 million mouths to feed, clothe, house, and educate each year. A 2-percent growth rate means a doubling time of thirty-five years. So in 35 years, at current rates, the world will have 8 billion people, with 160 million additions each year. And so on.

Clearly if growth is not curbed deliberately, then nature will do it for us through starvation, disease and war.

We can continue for a while to keep food production more or less abreast of people production. But eventually it *must* come to a halt. And yet even if reproduction were immediately reduced to the required just-over-two-children-per-family average worldwide, it would still take some seventy years before ZPG is reached (because so much of the world's population is now young). So we had better think ahead.

But although ZPG sounds like a basically simple idea, the implications and complications are staggeringly complex. In a free developed country, for instance, it would be almost impossible for a prosperous area to maintain ZPG even if the national rate of increase were held to zero; people would move there from less prosperous areas.

All industrialized economies have been geared to growth in numbers. We do not know what would happen in a true ZPG situation—where, for example, the ratio of older to younger persons increased and each person had to support a continually growing number of aged.

The answer may be that stagnation would set in, and stagnation is dullness, not utopia.

Rudolph Klein, research associate at London University, suggests that Great Britain might be taken as a model of the industrialized global-equilibrium society. He writes:

But if Britain is indeed to be taken as a model, then the prospect is gloomy. The experience of Britain would suggest that a non-growth society can produce as many and as unpleasant stresses on the social and political economy as industrial growth can impose on the ecological and natural resources of the globe. The stable-state advocates argue that growth does not guarantee greater social equality or justice. But the experience of Britain shows that stagnation (even under a Labour government ideologically sympathetic to equality) is no more helpful. Resentment of continuing inequalities is compounded by resentment of unemployment and of the failure of living standards to rise.[8]

But there is a greater problem How would ZPG be achieved? Shall we require a license? Allow a quota of one child to each person? Or perhaps two for each woman? Put a heavy tax on each successive child after two per family? Put a sterilizing potion in the drinking water, and allocate the antidote to legally acceptable persons?

There are many voluntary birth control programs in the less-developed countries, but in general they are not having much effect. At a recent World Food Conference concerning what could be done to help starving countries, not a word was officially spoken about birth control, not even the voluntary kind. No country is willing to be told what to do; many leaders shout "Genocide!" when the idea is mentioned. It is not hard to imagine what the response would be to a general call for one of the methods mentioned above.

In other words, everyone talks about the need for cutting population, but no one seems to have the vaguest notion of how this can be achieved, at least in free nations.* The result may be that we will, in desperation, turn to cutting economic growth just to show that something can be done. Yet this, without a concomitant cut in population growth, would be disastrous.

The general idea of ZEG is to stop us in our tracks. But most social programs have been financed out of growth funds, just as have most raises. In a no-growth society new social programs would have to fight for funding against established uses. And if it has been hard to correct the ills of society up to now, imagine what the situation would be in a no-growth society.

Suppose neither ZPG nor ZEG can be accomplished? Then that other alternative becomes tempting, namely to curb technology. That, surely, is one of the major objectives of many no-growth advocates, and it may be the place where, just to show we are doing something, cutting may begin. Yet, to me, this would be eliminating the one hope for a viable future.

As I see it, there are only three ways for the world to go:

1. That ZEG and ZPG will be imposed. The likelihood that this will happen while we maintain our material advantage—and have a stable world—is small. For that would mean ZEG for the world but NEG (negative economic growth) for us, at least until all countries had reached about the same level. For it is the gap that has

---

* Singapore has instituted some economic and social sanctions: families with more than two children are given lowered priority in government housing; the fourth child in a family is not permitted to go to the same school that the first three are attending; aid for childbirth costs decreases after the first two; and so on. But even though Singapore is small, has a high literacy rate, and is relatively well off, its growth rate is still not lower than the world average (2.2 percent). Yet that is considerably less than the 3- and even 4-percent growth rates in some less-developed countries.

made and continues to make for unhappiness, even among those who have enough to eat.

If this is the scenario for the future, then we in the developed countries may be lucky enough to have been alive at the peak of industrial civilization. Granted, luxury is not necessary for a satisfactory life. But for many of us with one house, one car, one television set, a couple of radios, central heating, one refrigerator, one washing machine, and one dryer, it doesn't really feel like too much. Perhaps it is. I for one would be sorry to see it go.

It is important to realize too that different groups will have different ideas about what is expendable. To some, having two cars is more important than symphonies, ballets, good books, and maybe even good food and health care. In other words, what *you* would like to keep may be considered nonessential by those in power, and vice versa.

2. That we will, by some deliberate plan, and worldwide cooperation, be able to provide everyone in the entire world with the wherewithal, or at least the option, to move up to a life of comfort as widely practiced in the industrial West. If this can be done without the accompanying problems of pollution and resource depletion, then all will be well—though of course only if population is stabilized at some point. But this scenario would require true international cooperation, including sharing of resources. It would also mean not spending several hundred billion dollars a year for arms. What we need, perhaps, is a modern-day Lysistrata.

Vast quantities of our material resources go into the so-called defense industries. Since it is not likely that these will be reduced, what we may well see is the specter of the defense group maintaining its power, while all other forms of technology are thinned. But if we talk of a 25-percent

cut in everything, then why can't we make that cut in the
military establishment right away?[9]

3. That things will somehow work out as outlined in
2 without the cooperation that at this moment appears to
be a hopeless dream. For the figures show a rather unexpected fact.[10] While the totals of resources and energy
use are indeed going up, the use per person levels off with
increasing income. It does not continue to go up.

Part of the reason has to do with what R. Buckminster
Fuller calls "ephemeralization"—doing more with less. His
example is that of the communications satellite, weighing
half a ton or so, which can accomplish what it otherwise
takes thousands of tons of transatlantic copper to do.
(Though he ignores the rockets and ground stations needed
to support the satellite, his point remains valid.) Wilfred
Malenbaum, professor of economics at the University of
Pennsylvania's Wharton School, figures that today we get
eight times more energy from coal than we did in 1900.[11]

On a per capita basis, developed countries consume
something like eight times more raw materials than less-
developed countries.[12] This would seem to indicate that as
the less-developed countries begin to develop, the materials
shortage problem will turn into total chaos. But, says
Professor Malenbaum, "The neo-Malthusians often fail to
recognize that advanced nations spend proportionately less
of each dollar of gross national product on raw materials
than do the developing countries." When the per capita income of a country rises to about $1,500 to $2,000 (1971
prices), it tends to spend less on materials and more on
services. The neo-Malthusians, he adds, "often fail to com-
pute the role of technology in developing new supplies or
wringing more from existing ones.[13]

Another important point is that birthrates in various

countries show a downward trend as gross national product per capita goes up.

The answer, then, seems inescapable. There is no hope worldwide for any change for the better, or for stability, unless and until gross national product per capita *in each country* rises to some reasonably high level, or at least until each country is as free of the natural scourges of mankind as the developed countries are today. All countries must either reach as comfortable a level as we have, or know that they can if they wish to. They must have that option, though they may decide not to take it.

Then and only then can we begin to talk honestly and reasonably of a worldwide cutting back, of global ZEG. It is hard to give up what you have; but it is easier than being told you can never have it.

Perhaps the world will realize that it is by no means necessary to go as far as we have in consumer goods production to have a satisfying life, especially if we can learn to curb our appetite. Our own experience seems to indicate that those who are willing to do with less are most often the better educated, the refugee from American abundance.

We must set an example. We must find a way of life that is comfortable but not wasteful. In Sweden people live quite comfortably, with probably the highest standard of living in Europe, yet use only half the energy per capita that we use. They tend to live more in apartments, however, and often in conditions that we would consider crowded. Nevertheless, there are vast areas for economizing in our country in both energy and materials expenditure; it may hurt a bit, but we had better pitch in. Perhaps we can learn to direct our consumer spending away from goods (or "bads," as one economist calls them) and toward better services in health, education, and leisure.

ZTG?

Technological growth and economic growth are not the same; indeed they are not even inextricably coupled. It is quite possible to conceive of economic growth continuing with no change in technology. It would only be necessary to increase the inputs—which would mean, of course, a concomitant increase in resource use and pollution.*

It is similarly possible to conceive of a no-economic-growth situation, in which technological change continues. The result of such a case could well be ZEG with *reduced* input of energy and raw materials!

Economic growth *uses* resources; technology *creates* new ones: coal for wood, oil for coal, uranium for oil, and maybe solar power for uranium.

Trying to control pollution and raw materials use by controlling growth, therefore, is like trying to make someone quit smoking by limiting his income. The connection is just too indirect, too vague. Controlling growth would not stop production and use of dangerous chemicals, nor the production of unpleasant effluent and its introduction into our environment. Without stringent regulation, pollution might even increase—if, for example, a company has been financing pollution control out of growth funds.

As Marc J. Roberts put it, "The way to control pollution is to control pollution, not growth."[14] The way to cut resource use is to cut resource use.

We must move away from the throwaway or cowboy mentality and toward the spaceship idea of reusing and conserving in every possible way. Deliberately designing

---

* An economist might argue that this is not real economic growth—that real economic growth can only come from increases in productivity, a large portion of which are due to new technology. The point remains valid in either case.

things to last longer and to be more easily repairable is one way. Cutting down on yearly model changes is another.

It is tempting to say that this cannot be done in a capitalist, profit-oriented economy like ours. But who would have imagined that profit-oriented companies and public utilities that, only a decade ago, were exerting every effort to increase demand for their electricity or fuel would today be studying "demarketing" procedures designed to cut use of their goods and services? [15]

Nevertheless, our society must go much further than this. We must find a way to decrease "Jonesism," [16] the wasteful practice of keeping up with the Joneses—which is practiced by corporations and even countries, as well as individuals, especially in the United States. Changing people's habits is hard, but we had better find a way, or continued deterioration will force us to give up many of the freedoms we cherish, without accomplishing our goal.

Until fairly recently increases in wages have been pretty well matched by increased productivity—usually due to mechanization or more efficient methods of production, rarely to harder or faster work by labor. For various reasons this is no longer true, and is surely one important cause of the inflation we are suffering today.[17]

We are competing in a world market. The reason we have been able to stay ahead is our still-fantastic productivity. An American coal miner can produce several times more than a British one. But this has little to do with the miner; it is mainly because of mechanization.

Other countries, with lower labor costs, still manage to compete very well, however. And as other countries catch up, our edge will decrease and could be lost. The situation was clearly stated by the Council of Economic Advisors back in 1964: "Within a few years after the introduction of almost any new product in today's world, a dozen nations will be able to compete with us in its production. To maintain or expand our share of world exports we must continually be in the vanguard of product development." [18]

We can put import duties on goods that other countries can produce more cheaply: we can "buy American" rather than importing Spanish shoes, Japanese cameras and German cars. That works for a while, and it does help protect certain American manufacturers, but it is obviously not a satisfactory solution.

The point of course is that we are not an island unto ourselves; we are not self-sufficient—not politically, not economically, and certainly not in terms of raw materials. We import enormous amounts of materials, including great quantities of crude oil. We *must* export an equivalent or higher amount of goods and services if we are not to become a debtor nation.

One of our major exports is technology! This can be in the form of (1) technologically advanced goods, such as the Boeing 747 or enriched uranium, (2) direct foreign investment (building plants and producing goods overseas); or (3) licensing or royalty arrangements for the right to use one of our methods. Clearly a country in the vanguard of technology will hold an economic advantage.

The difference between what a country pays out and what it takes in for "knowhow" is called the "technological balance of payments." For a long time now the balance has been very much in our favor. Figures for 1972 show that we took in more than eight times what we paid out in this area.[19]

But while our technological balance of payments has been holding up well, Dr. Michael Boretsky, senior policy analyst with the U.S. Department of Commerce, argues that this is only because we are exporting more of this kind of technology—he calls it *naked technology*—than we should be. Indeed, he feels this is a major reason for our declining trade position in the last few years, in that it has permitted the technological and industrial capabilities of other countries, particularly Japan and those of Western Europe, to grow faster than they have here. In other words, we are losing our competitive edge. After many years of comfortable trade surplus, in 1971 and 1972 (before the oil crisis) we had negative trade balances of $1.5 billion and $5.8 billion.[20]

Although the United States continues preeminent in the world today in these areas, Boretsky points out: "Toward the end of the 1960s and especially since 1971 . . . a number of societal attitudes toward technology, on the one hand, and quite a few unfavorable economic trends in technology on the other, have induced considerable speculation about America's economic and political future. . . . The most far-reaching among these problems are the decline in productivity growth, both of labor and capital, and the deterioration in the U.S. foreign trade position."[21] That is to say, output per man hour has since the middle 1960s been growing at only half the rate of the previous century, while productivity gains have advanced well beyond ours in other countries.

An OECD report dating back to the mid-1960s showed that even then in several European countries (Belgium, France, Germany, the Netherlands) a higher proportion of students were getting degrees in technology than was true here. In Great Britain a student was twice as likely to take a first degree in engineering as he would be here! Today, if anything, the differences are probably even higher.

We sometimes think of ourselves as a highly industrialized country (which is true), producing vast numbers of scientists and engineers (which is only partly true), and of other countries as interested only in art, culture, and the good things in life (which is no more true of them than it is of us).

"So what if we are not rich and powerful and advanced?" the growth critics ask.

There are plenty of selfish reasons that can be given in answer. But there are unselfish ones, too. Consider that we are on the one hand being exhorted to cut down, and on the other to help feed, train, and equip many of the third-world countries desperately in need of help. Where is the capital for all of this to come from? From where, if not from the technologically advanced countries, will the breakthroughs needed for solar power and clean nuclear energy come?

An old Chinese proverb says that if you give a man a fish you feed him for a day; if you teach him to fish you feed him for life.

We must therefore be very careful when we talk of curbing growth, particularly technological growth. In 1970 a statement was issued at the close of a conference held under the joint sponsorship of the Aspen Institute for Humanistic Studies and the International Association for Cultural Freedom which seems to sum up the situation rather well:

> The need is not for the slower development of technology, either in advanced or in developing countries. Such a slowdown would cruelly sacrifice the interests of millions of underprivileged people whose hopes and expectations cannot begin to be met without more technology. The need is rather for more thoughtful and careful application of new technologies to prevent both long range damage to the earth and violence to

human values and to foster social, economic and cultural development.[22]

How we might at least hope to do this is spelled out in the next two chapters.

9 Technology and the Citizen

Now, all of a sudden, people have awakened to the fact
that science and technology are just the latest expression
of power and that those who control them have become the
new bosses, exactly as the feudal landlords who owned
the land, or the capitalist pioneers who owned the factories,
became the bosses of earlier generations. Ordinary people
will not now be satisfied until they have got their hands
on this power and have turned it to meet their needs.
—*Anthony Wedgewood Benn*

What if someone could have foreseen right at the beginning the congestion, asphaltization, energy shortage, pollution, and other problems associated with the automobile?
Would things have been different today?

The question cannot be answered with any surety, for
our forebears were no better at looking into the future
than we have been. We do know, though, that after some
early objections there was a large and rapidly growing
feeling that the motor car was a positive development,
one that solved all sorts of quite real problems. In 1899,
the magazine *Scientific American* said of it: "In reality,
there is no mechanism more inoffensive, no means of
transportation more sure and safe." [1]

My guess is that even if a true, believable, and believed prediction of the future of the automobile had been made, our predecessors would have shrugged their shoulders and said, "Does it make sense for us to worry about what's going to happen in the future? Who knows, maybe the electric car will be perfected by then, or antigravity discovered, or whatever. I say let's go ahead with the motor car. Are future generations worried about our flies, disease, and dead horses? Of course not. Why should we worry about them?"

Does this mean we should not try to forecast and assess the future results of the technology being developed today? Not at all. For one thing, times have changed. Industrial society is ready for such an attempt, which was not true at the turn of the century. That was a time of great optimism. Science and technology were riding high, and there was still a belief in automatic progress. Indeed, technology had become a byword of progress; and big business, which thrived on a healthy diet of technology, was admired in a manner that would be incomprehensible to most of us today. William and Henry James are said to have been embarrassed throughout much of their early lives because their father was an intellectual and not a businessman like their friends' fathers.

Today, as we have seen, there is a widespread suspicion of science and technology, indeed of all institutions. And this may be a good sign, for perhaps we *have* been like the person with a new electric drill, and have tried to use it to do everything. And when it has failed, as had to happen, we became angry.

The result of the change may be that technology can, for the first time, be looked at sensibly, with realistic expectations, and that there is as a result a greater likelihood of our being able to forecast what the unexpected outcomes of a new, and supposedly better, technology may be.

Technology Assessment

Eric Hoffer points out: "No other generation has been made so poignantly conscious of the perils of doing good. We know that to set out to do good is to run the gauntlet of baffling, grotesque side effects."[2] Still, we must try. One of the more recent attempts is called *technology assessment*.

Almost any study of a new proposal—a dam, airport expansion, the supersonic transport—is a kind of technology assessment. The practice is not new; as long ago as the late 1930s the Department of Agriculture studied the implications of the mechanization of cotton growing and harvesting. They clearly predicted the displacement of large numbers of agricultural workers from the South. Policy makers, however, totally ignored the study, with results that are well known.*

In 1970 the National Environmental Policy Act required that Federal Agencies include "environmental impact statements" in all proposals for new techniques or structures that might have an impact on the environment. This was a step in the right direction, but limited to environmental considerations.

In 1972 there was set up in Congress the Office of Technology Assessment (OTA),† which is concerned with the

---

* There is another aspect to the matter. Mechanization is blamed for the plight of the black today. But a case can be made for the idea that we are really paying for several centuries of slavery, and that if there had been machines to do the work the slave was dragged off to do, as there is now, slavery would not have paid off, and it would not, therefore, have existed.

† OTA should not be confused with the Department of Commerce's Office of Technology Assessment and Forecast, and the National Science Foundation's Office of R & D Assessment, each of which, in its own way, serves a different function.

physical, biological, economic, social, and political effects of selected new technologies.

The OTA, then, is a way for Congress to obtain information on any new technologies that they are being asked to fund. Because the government supports a great deal of research and development, it will be able to exert a great deal of leverage. If the dangers and disadvantages appear to outweigh the positive effects, funding may not be given, which could effectively eliminate the development.

The OTA director and staff are responsible to a board of directors made up of six members each of the Senate and House of Representatives. Technical and public input are provided by a council made up mainly of representatives from education and industry. Assessments are done on contract by universities, nonprofit and private research organizations, and ad hoc groups. Work is initiated on the request of heads of various Congressional committees. The first areas chosen to be worked on reflect the problems of our times: energy, food, health, transportation, oceans, materials, world trade (including problems of international patents), and pharmaceutical drug bioequivalency.

As an example, a half-million-dollar contract has been given to a private firm to evaluate the potential risks, benefits and consequences of oil and gas development in the waters off the Delaware–New Jersey coast. The resulting study will advise Congress on proposed widespread drilling for oil and natural gas in the Outer Continental Shelf area known as the Baltimore Canyon; on alternative methods of recovery of these resources if they are found; and on how exploration and production—if considered advisable—might be undertaken with the least possible harm to the environment. Two other, related technologies are being included in the study: deepwater ports for supertankers and offshore siting for nuclear power plants.

A local input will be supplied to the research company by an eleven-member advisory committee recruited from the New Jersey–Delaware area.

OTA is expected to be more than simply an "early warning system"; it must play a positive role as well, looking for positive opportunities in technology and calling them as well as the dangers to the attention of Congress. But it is not a decision-making body; that role is left to Congress.

The fact that such an office has been set up by no means ensures clear sailing, of course. The assessors are, after all, human. And humans have wants, needs, desires —and obligations. One of the first acts of the Interstate Commerce Commission, which was created in 1887, was to call for rails made of iron rather than steel, even though steel was better all around. Reason? The iron was being made domestically; steel for rails still had to be imported.

Another example of human fallibility: the German leaders during World War II are reported to have decided that radar, the basic principles of which were known in Germany, was strictly a defensive weapon; since they were going to win on the basis of a quick, overwhelmingly powerful offensive, there was no need to devote resources to developing this new electronic device. By the time they woke up to the realities of the situation, it was too late.

It is therefore important to understand that the assessments will not be, cannot be, "gospel." They cannot be more than guesstimates, first-order approximations to future truth.

The Public Interest

Assessments, indeed any consideration of whether to provide financial and/or moral support for technology, must be based on some criterion or criteria. The criterion we hear most often today is the "public interest." If it does not provide a net gain to the public, it should be scrapped.

But which public? What is good for one may be anathema to another. Even the idea of "the greatest good for the greatest number" is not a complete answer, for that could mean the misery and tyranny of one or more minority groups. At the beginning we shall probably have to be satisfied with determining those things we *don't* want.

Even in terms of something so obvious as the excessive amount of packaging we use in the United States, the options are not always clear. For example, ask the small dry goods or clothing retailer whether he is willing to give up plastic packaging, to go back to having his white goods turn dusty and gray before he can sell them.

If we were to decide to stop or cut back all technological work that does not have "social relevance" until such time as all present social problems have been solved, we might end up, as we saw in the last chapter, not only with the same problems but with a large net loss.

Do we eliminate monuments such as the Lincoln and Jefferson memorials? We might end up making our basic civilization and culture much poorer.

If the motto, "Not a sparrow shall fall," were really to be observed, then a lot of very important work—including schools, sewers, subways, and burying of utility lines—might never get done.

We must keep clear the distinction between goals and means. It is all very well to say we should plan, we should regulate, we should control. It is manifestly clear that there is much to be done. But things are happening. In the area of pollution and poisons in the environment, for example, Harvard's Harvey Brooks predicts that a national law will shortly be passed requiring a prior assessment before any new chemical is introduced into the environment.

In resource use there are some obvious areas of waste —huge cars, yachts, and mansions; uninsulated and poorly planned homes; and others we have mentioned. But in the main we haven't yet developed the tools, the data, the techniques to be sure that the means chosen will carry us to our ends.

Consider the "cancer crusade" launched by the Nixon administration. Cancer research was given a boost in funding while the funding of many other groups was cut down or eliminated altogether. Who could argue against "a humanitarian" goal like wiping out cancer? The problem, pointed out by noncancer researchers mostly, was that the field was not ripe for such a crusade. Not enough basic, wandering, undirected research had been done; too few of the fundamental problems had been solved. One cannot, they pointed out, handle this like the moon shot, which was, after all, an engineering problem, and a good example of what is meant by "mission-oriented" or goal-directed research. It can work in some areas, such as the crash development of synthetic rubber during World War II, or the development of atomic energy, or the moon shot, but not in others where the required path, the approach, is not clearly defined.

Oddly enough, the scientific community, because of its greater awareness of what the effects might be *in the future*, may end up being the most conservative element in our technological society. Who but a scientist could have come up with the idea that the ozone layer in our atmosphere is threatened by our activities (e.g., SST, spray cans, etc.)?[3] It happens often that the technical people are the "conservatives": in 1962 a statement was issued by seventy-five top mathematicians decrying the "new math" (which is now losing its strong position). But fashion and fad seemingly must be given their chance in our innovative society.

Directed Research

Jacques Barzun one called science "that glorious entertainment." Now people working in these fields are being called on to pay for their play. Those who provide the funds are saying, "If you want money and equipment, you'll have to give us very good reasons for it, including telling us how what you do will be good for society. As a matter of fact, I don't think what you are working on is really very important; why don't you switch over to . . ."

Researchers are understandably alarmed: "If you tell me what to do, I may end up not working on what I enjoy doing and may be very much less productive. The very idea we need may never come forth, or may be long delayed because of this idea of directed research. Suppose a group had been put in charge of doing something about polio. We might very well have gotten improved iron lungs and braces, rather than a vaccine."

A vast number of modern approaches to, and devices for, saving lives and alleviating pain have come out of totally unrelated developments. Smaller, more powerful batteries: who would have thought that these would be useful in medicine? Today they are widely used in pacemakers that help about 125,000 people whose own heartbeat regulators do not function properly. Nuclear techniques, with their high concentration of energy, are also being used in pacemakers, and may one day be used to power sophisticated, computer-controlled prostheses. Who would have thought that experiments in static electricity in previous centuries would one day lead to electrostatic precipitators that clean smoke coming out of power plant chimneys?

Fortunately, the membership of OTA is diverse, involving representatives from many different groups, and so there is hope that they can avoid the trap of thinking they know more than they do.

Citizen Participation

Until recently it was considered pointless for the untrained person to even try to comprehend what is going on. "Just stay out of our way" was more or less the attitude of the researchers. "We'll take care of your needs (eventually)."

But a real change is taking place. More and more, the public as well as the funding agencies are asking questions.

The kinds of questions being asked are changing too. We hear not only "Will it work?" or "Will it make money?" or even "Will it help us do X faster, cheaper, or more easily?" but also, "What effect will this have on our environment?" and "Will it in the long run do more harm than good?"

Even among the individual scientists and engineers themselves, some surprisingly incisive questioning is taking place.[4]

Questions about the dangers or even advisability of such technologies as space or nuclear power were traditionally left to the agencies that specialized in them. But recently the leaders of the American Physical Society (APS), decided to go ahead with an APS-sponsored study of the technical aspects of the safety of nuclear reactors; it is also looking into the possibility of advancing the public-interest process by creating what has been named the Public Interest Science Clearinghouse to bring together technical people and the organizations that need them. A conference was held on "Scientists in the Public Interest: the Role of Professional Societies," and a new *Journal of Physics and Society* is being considered.*

Certainly the decision-making, or at least the advisory, process is moving down the ladder.[5] No longer are organizations like the AEC and NASA able to operate with the freedom they have traditionally had.

Indeed the very idea of such organizations is being questioned. That is, was it right for a group like the AEC to have under its aegis both the regulatory and developmental aspects? It seems clear (in hindsight) that the regulatory aspects will bend to the promotional when there is conflict, and there has been plenty of that in the nuclear field recently.

Granted, the AEC came up with a going nuclear capability in the electric power field. So perhaps the combined approach was best in the early years; but now the time has arrived for change. The AEC has been split up into two agencies, one of them responsible for research in the overall energy field (Energy Research and Development Administration), and the other for regulation of nuclear energy (Nuclear Regulatory Commission). This should mean that other energy forms will have a better chance at competing for research funds. Combined with what seems to be a growing public resistance to nuclear technology and other problems, it could even mean a significant change in direction. Daniel S. Greenberg, a knowledgeable

---

* This is not all pure altruism, of course. Physicists are worried about the decline in interest of the student body in their subject, and in loss of support among government, university administration, and the public. In one meeting on the subject a housewife suggested, "Get rid of the name physics!"

reporter of the science scene, asks: "Is it possible that atomic power is going to experience the fate of the dirigible, and fade into history as a once-promising technology that just didn't make it?" [6] Again, public input is having an effect.

Many of the old-line groups, including the National Science Foundation, the National Academy of Sciences, and its offshoot the National Academy of Engineering, are getting more deeply involved in the area of public responsibility. A number of new groups have also formed in recent years, including the Academy for Contemporary Problems (Columbus, Ohio); Center for Science in the Public Interest (Washington, D.C.); Committee for Social Responsibility in Engineering (New York City); Institute of Society, Ethics, and the Life Sciences (Hastings-on-Hudson, N.Y.); Institute on Man and Science (Rensselaerville, N.Y.); Scientists and Engineers for Social and Political Action (Jamaica Plains, Mass.); Scientists' Institute for Public Information (New York City); National Council for the Public Assessment of Technology (Washington, D.C.); Union of Concerned Scientists (Cambridge, Mass.); and Federation of American Scientists (Washington, D.C.).

Many of these groups put out publications concerned with the general topic of society, science, and technology, and would welcome support in the form of subscriptions donations and perhaps even volunteer labor. The Cornell University "Guide to the Field" lists 168 publications that are in some way involved in this broad field.

Lay people are also getting more directly involved through such conservation and environmental groups as the Sierra Club (San Francisco and many branches), Friends of the Earth (San Francisco), Citizens for Clean Air (New York City), National Audubon Society (New

York City), Colorado Open Space Council (Denver, Colo.), and the National Wildlife Federation (Washington, D.C.).

There seems to be a greater recognition of, and acceptance of, the fact that the layman can play a part in the setting of policy. During the spring and summer of 1974, the new Federal Energy Administration held a series of some twenty public hearings across the United States, some of them lasting a full week, on Project Independence—the attempt on the part of the United States to make itself energy independent over the next decade.

Role-playing experiments are in progress in various colleges and universities, with students playing the part of concerned citizens whose interests—as homeowners, conservationists, business people, or what have you—are at stake.

The OTA itself has begun to exchange information with the public by means of informal workshop meetings and questionnaires. In an invitation concerning the New Jersey–Delaware Offshore Energy and Coastal Zone Assessment, it was pointed out that, "among those being invited to the workshop are citizens from the industrial, commercial, labor, consumer, environmental and public interest communities."

The objective of this input, says the OTA, "is to help insure that the final assessment report does not omit or neglect factors within the scope of the assessment which citizens consider relevant and important." [7]

Citizens can make their influence felt in other ways as well. A fairly new way is through litigation: the courts have now decided that one need not be actually hurt by a move before being permitted to bring the matter to the courts. Specialized nonprofit public-interest law organizations such as the Environmental Defense Fund, the National Resources Defense Council, and the Center for Law in the Public Interest have come into being as a result.

Other options include citizen's petitions to state and national legislatures; voting, writing, or speaking to elected representatives, taking part in public opinion or media polls when requested to do so, forming or joining community, state, or national action groups, letters to the media, and testimony at public hearings.[8]

A single action along any of these lines is not likely to have great effect. But it can start a movement, or add a little bit to it, or start a congressman thinking, which *can* lead to change. A large-scale effort that seems to have finally put a clamp on widespread encroachment into San Francisco Bay by developers is said to have begun with the efforts of "three dowagers and a disc jockey." [9]

There are many examples of projects that have, in recent years, been postponed or turned aside because of the efforts of conservation, environmental and other public groups. The question of how useful a development will be to society as a whole *can* have an effect on whether a development is carried forward. This was a major factor in the cancellation of the SST (supersonic transport) in 1971. That is to say, relatively few people would have gained, while many would have suffered its effects. Similarly, a plan for expansion of New York's Kennedy Airport into Jamaica Bay was turned down because of strong opposition from public and community groups, in spite of a strong belief at the time that additional capacity was desperately needed.

More recently, the Army Corps of Engineers, which for

a long time pretty much had its way whenever it wanted to build, was prevented—after a thirteen-year battle!—from constructing a dam across the Delaware River at Tocks Island, north of Stroudsburg, Pennsylvania. Said *The New York Times:*

> Probably enough paper has gone into studies, reports, briefs, and controversial literature to make a dam all by itself . . .
>
> Yet opponents of the dam—mostly ad hoc environmental groups and outraged people in the affected communities—refused to give up. Like the defeat of the Florida Barge Canal a few years ago, the ultimate saving of the Delaware River is an encouraging demonstration of the continuing power of the people to protect their environment when they have the will to stand up for it.[10]

Nuclear plants are being held up by citizens' protests in New Jersey, Kansas, Missouri, and elsewhere. And a virtual one-man campaign has caused one of the largest and most unwieldy of all institutions, the Postal Service, to reverse a policy. After five years of using only the state name (abbreviated) and the last three digits of the zip code, rather than the city name, on postmarks, the Postal Service has gone back to using the name of the processing center where the letter is postmarked.

Ray Geiger, owner of the *Farmers' Almanac*, is the man who did it. He started with an article, expanded his campaign to radio and television—a grand total of 325 appearances!—and urged members of Congress to join in. About 150 of them eventually did.[11]

All of these methods and organizations imply external pressure on the "wrong-doers." To these must be added what Ralph Nader, the country's leading consumer advocate, calls "whistle blowing." This is the action taken by someone with inside knowledge of undesirable actions by an organization of which he or she is a part. One example is that of Dr. Jacqueline Verrett, a biochemist working for the FDA, who blew the whistle on cyclamates, to the embarrassment of the FDA. The result, interestingly, was first censure and then a promotion.*

And Edward Gregory, an employee of General Motors, was instrumental in what turned out to be the largest recall of automobiles in history—2.4 million Chevrolets, called back because of faulty exhaust systems. He had to fight his own management all the way. Happily, he is still employed by the company.†

Other examples have not ended so happily.[12] Some people have not only lost their jobs, but have even been blacklisted in their fields, though of course this is denied. Nader says that blue-collar workers have some protection from their unions, while scientists have very little. He suggests that some sort of protection should be built into our laws to protect anyone brave enough to speak up like this. A group called the Clearinghouse for Professional Responsibility, in Washington, D.C., has been helping several university scientists who may have been denied tenure for speaking out in this way.[13]

---

* Cyclamates were banned as a food additive in 1969 on the grounds that they are carcinogenic, or cancer producing. Abbott Laboratories, believing the evidence was not conclusive, has petitioned the FDA to allow the return of cyclamates to the market. The question has been referred to the National Cancer Institute, and at this time (August 1975) remains unresolved. In the meantime, reports the *FDA Consumer*, "Tests in animals have raised questions about a possible association between cyclamates and testicular atrophy as well as about possible adverse effects on the cardiovascular system. These matters require additional study" (April 1975, p. 1).

† When a scientist goes outside the usual governmental or industrial channels to help bring issues to the public or to the courts, the expression "public interest science" is also sometimes used.

Consumer Action

And with all this we have not yet mentioned one of the most important ways the public has of making its voice heard—in the role of consumer. Certainly when we begin to consume somewhat less, or to buy more sensibly, then fewer, and less wasteful, products will be developed and produced.

Forget about electric can openers; use windows instead of air conditioners on cooler days; turn off lights when not in use; share cars; rediscover bikes and feet; use white instead of dyed paper goods (the dyes are poisonous and difficult to handle in our waste systems).

The possibilities, when once you begin to think in these terms, are endless. Perhaps the main advance will take place when we stop giving in to the style changers. People throw out perfectly good and even handsome clothing in a desperate effort to keep up with designers, but they are chasing a moving target and will never catch up. The same holds, though to a lesser extent, with cars, appliances, and a host of other products.

Our cars have become so complicated that few can still be worked on by the motorist. The energy situation may be a blessing in disguise; it may force us to get rid of all the extras—air conditioning, power steering and windows, automatic transmission, big engines, and so on. We may even end up with the electric car, which, though an example of advanced technology, is much simpler than the internal combustion system.

But consumer action needs some understanding, some depth of knowledge. In our day of energy shortage, the laws of thermodynamics, including the law of conservation of energy, would surely help the average person find his own ways of saving fuel and electricity.

Yet many of us are permitted to go through our entire high school and college career without taking a single course in science or technology. We are not even given the technological equivalent of art or music appreciation.

Indeed, a curious inversion takes place in our education process. In high school, English, art, history, and politics (the latter two often lumped into "social studies") are considered important enough to be required subjects, while physics, chemistry, biology, and technology are not. Thus if a student is, for any reason, frightened by or is otherwise uncomfortable with any of these latter areas, he is as a result likely to remain so for the rest of his life. Yet in later life we find that science is, or seems to be, more important than art or history, for it is supported by large quantities of money. That money is supplied, ultimately, by the taxpayer, who as a result of his early deprivation, has little idea of what (s)he is being asked to support. There is support for other fields—the arts, history, English, and so on—but it is far smaller.

Public Service

The matter of support brings up an interesting possibility, namely inculcation of the idea of public service into our population. And a good place to start is with those who are being educated with public money. Many people training in medicine and science are being supported by public funds, and should be required to donate some part of their time to public service. The National Research Act, passed in 1974 and implemented in 1975, is a step in the right direction, for it requires that recipients of training grants and fellowships in science and medicine provide a given number of years (depending on the time of support and the area of public support) in education, public health service, work in doctor-poor areas, and so on.[14]

But this can be carried much further, perhaps even including teaching or counseling in the lower grades, acting as science club leaders, and so on. This would have the advantage of making the teaching of science and technology more exciting.

It was found in the Peace Corps that what the developing countries needed most of all was help in basic technology, not good-natured young idealists who could teach music or English but knew nothing about sanitation, agriculture, transportation urban planning, health care, or even building a latrine.*

In the United States, shortages are predicted in technical personnel for such areas as mining, agriculture, and public health.

My own feeling, clearly, is that more science, better taught, would be a highly desirable approach.[15] But, as usual, not everyone agrees. I once asked Dean Don K. Price of Harvard whether he thought it might not be a good idea to require more science and technology in university education to help people cope with this highly technological world.

Here is his answer: "It seems to me the notion that we can make everybody into a scientist is rather utopian, and I'm not even sure it would be a good thing. But it is possible to have the average citizen much more literate about the limitations of science and the ways in which it ties in with the political and administrative problems,

---

* Of course, English might be useful in that it could eventually give access to technical books, articles, and so on, but the immediate need is for direct technical assistance.

and to quit teaching them the wrong things, that is, how science is either an engine of automatic progress on the one hand, or a devil that we've got to abolish, on the other."

He added: "You can do a lot better job in educating people in the ways in which science affects politics than you can in educating them in the intricacies of the basic physical sciences which are too tough and mathematical for most people, such as calculus and nuclear physics. And my experience is that that isn't what you need to know to control science. . . . The issues that get to the top of the government hierarchy and that are of most interest to the citizens are not typically issues that the scientist as such can answer conclusively. Besides, a biologist doesn't necessarily know much more about nuclear physics than does a student of the classics."

Well, maybe required courses in science and technology would be going too far. But they should certainly be stressed more than they have been. Otherwise we will continue to have the kind of situation described by Dr. Dixy Lee Ray, past chairman of the now-abolished Atomic Energy Commission. At a congressional hearing she attended she was greatly impressed by the intent and good will of the legislators on the relevant committees. "And yet," she pointed out,

> when it came to scientific questions—and there were some of the finest scientific minds of the century sitting at the witness tables—[the two sides] were quite unable to understand each other. This led to frustration on both sides, which led to irritation. The scientists tended after a while to become condescending, and the Congressmen impatient." [16]

Better comprehension of science and technology would

help prevent this. It would also prevent the wholesale swallowing of foolish, irresponsible, exaggerated, and even outlandish statements by scientists and journalists which, as we have seen, tend to create all kinds of unrealistic expectations and fears in the eyes and minds of the public.

What we also need, as Prof. Joseph Weizenbaum of MIT points out, "is something that's very hard to get; what we need is very, very large doses of critical thinking, and that's the one thing most people are not being trained to do. We still need a hell of a lot of incisive question asking." [17]

No one is suggesting that politicians must become scientists. They are generally different sorts of people. Indeed, the scientific parliamentarian may not be a good idea at all. The story is told of the great Isaac Newton, who was a member of the British parliament for a while. He spoke only once, it is said, and that was to ask that a window be closed.

We need thinkers, we need doers, and we need dreamers. But all of these representatives of the human race need to be able to communicate with the others.

Even if we do not manage to make lay people more literate in a scientific sense, their greater interest today in what is going on may prevent some of the foolishness we have seen in the past.

State or Federal Control?

We are facing a paradox, however: the paradox of the people who continually call for the government to step in and regulate the price of oil and gas—to keep big business in line, that is—yet who complain of big government and even government interference in their lives.[18] The dilemma is not new, however. It was faced, for example, by Thomas Jefferson. Almost exactly as we hear today the complaints of bigness, of excessive growth, so too did Jefferson wish to hold down economic development, to keep the United States an agrarian country. But to do so would have required precisely the kind of governmental power he detested.

So we will undoubtedly continue to be faced with dilemma after dilemma. We may not be able to predict *anything* accurately. But it seems that from now on no movement to any major system will take place without hard, suspicious, widespread questioning. This may be the most important change that has taken place in our society since the Industrial Revolution began.

10 Sociotechnological Needs

The stoical scheme of supplying our wants by lopping off
our desires is like cutting off our feet when we want shoes.
—*Jonathan Swift*

Probably the greatest bar to innovative thinking in the
energy field has been, until recently, the ready availability
of low-cost petroleum and, to a lesser extent, coal and
natural gas. The result is that, during the last century,
a period of extraordinary technological fecundity, we have
not managed to bring to fruition a single satisfactory
alternative, despite a large number of possible alternatives.

True, the French built an electric generating plant that
uses the rise and fall of the tides. And there have been
some limited applications of solar power for hot water
and space heat in Israel, Australia, Japan, and elsewhere;
earth heat is used for similar purposes and even for
production of electricity in those areas where hot water
or steam is readily available, as in Italy, Australia, Iceland,
and our own West Coast. But in all of the other countries
the cost of oil has always been much higher than it has
been here.

While the immediate problem in oil is the heavy dependence of the developed countries on Middle East crude, the real issue goes much deeper. Basically, it boils down to the fact that we have been ripping through a storehouse that took nature millions of years to put together. Or, as the pioneer French futurist Bertrand de Jouvenal, puts it, "We have set about procuring energy much as soldiers in winter might obtain heat by burning a wooden house." [1]

Further, the more easily obtainable fossil fuels—natural gas, coal and oil—are being taken out first; this means that not only the cost of obtaining fuel must rise, but that the energy used in obtaining it will also rise, eventually to the point where we are using almost as much energy as we are obtaining. At that point we approach the apex of diminishing returns.[2]

We are, in other words, living off our capital.

Estimates of present oil and natural gas reserves, at current and projected use rates, put the time left at a few decades!

And although the United States has been blessed with several centuries' worth of coal at present rates of use, we have not found a way to use it without serious pollution of the air and numerous other problems. Strip or surface mining does great damage to the landscape; deep mining, while safer than it used to be, remains one of the most dangerous occupations in the country.[3] It is also labor intensive, which means high production costs.

Although we have plenty of oil shale, we have not found a way to extract the oil from this rocky material at a reasonable price and without serious environmental problems; for every barrel of oil obtained, well over a ton of rock must be handled and disposed of.

So we now find ourselves with nuclear power as the only

real, established alternative to our present dependence on fossil fuels. But nuclear power presents a number of still unsolved major problems. There is as yet no good way to dispose of the radioactive waste material, no assurance of complete containment in case of accident, and always the possibility of fissionable material falling into the wrong hands.

The next half-step beyond present fission reactors is the breeder reactor. This form has one special advantage: it produces more fuel than it consumes. But the fuel is plutonium—one of the most toxic substances known to man.

The nuclear industry is far behind in its construction schedule; this is partly due to management problems, partly to site approval procedures, and partly to technical difficulties. The delay, say some of its critics, may be a blessing in disguise.

Of course, we are speaking now with the advantage of hindsight, but it does seem that the decision to go for nuclear power put all our development eggs in one basket. Although we were smart enough to realize that complete dependence on fossil fuels was a poor idea (the first NSF budget in 1950 included a small amount for solar energy research!), the idea of developing a number of other sources never took hold. Perhaps it was too complex to contemplate.

Today we are desperately seeking new forms of energy, trying to do on a crash basis what should have been done at leisure over the past several decades. Unfortunately, every alternative source presently offered as the panacea suffers from one or more of a variety of limitations, including high cost (e.g., production of synthetic oil, gasoline or gas from coal), limited potential (tides), unreliability and variability with time of day or year (wind, waves, and solar), various potential dangers (nuclear), high investment in equipment (most), unproven (fusion, ocean thermal energy).

The use of solar energy to heat a house, for instance, requires some way to store the heat for nighttime or cloudy periods, which usually involves multithousand gallon water tanks, or large volumes of concrete, stone, or iron. Under severe conditions, the storage would be good for only about three days, which means a backup system is required. But a backup system means adding more expense to an already costly installation.[4] With wind power and solar power used to generate electricity, there is currently *no* good way to store excess electricity.

Not a single alternative source boasts the compactness, convenience, flexibility, and low production price of petroleum, even at today's inflated prices. We have been hearing about electric cars and trucks for decades; but after a strong early start they fell behind the internal combustion engine and never caught up again.

The fuel cell, a batterylike device that directly converts fuel to electricity, has great potential, promising as much as 150 miles per gallon. Yet though the concept was invented in the nineteenth century, the equipment is still complicated, expensive, and unreliable.

Methanol (methyl alcohol) and ethanol (ethyl alcohol) are clean-burning liquid fuels that have potential use in automobiles. Both can be made from wood or agricultural waste products, which makes them renewable.[5] But the energy content per gallon is much lower than in gasoline, and they are still considerably more expensive to produce. Ethanol is actually grain alcohol, the basic constituent of hard liquor. Methanol is the better fuel, but is harder to make from vegetation; it can be manufactured from fossil fuels, however. This wouldn't help the shortage problem, though it does decrease pollution.

John McPhee maintains that "roughly a thirtieth of the annual growth of all the trees on earth could yield alcohol enough to run everything that now uses coal and petroleum—every airplane, every industry, every automobile." [6] In the automobile, advantages claimed for alcohols are better fuel economy,[7] lower exhaust emissions, better performance, and the ability to be mixed with gasoline. But E. E. Wigg, senior research chemist at Exxon, counters that these are only significant for the older cars with richer carburation, the cars that are rapidly disappearing from the roads.[8]

Hydrogen too has been put forth as an energy panacea. It is essentially inexhaustible—nuclear, wind, or solar energy, perhaps in the form of photosynthesis, can be used to split water into oxygen and hydrogen—and it makes a perfectly clean fuel: as it burns it combines with oxygen to form water again. But there are as yet no economical processes by which to make it, nor is there a reasonable way to store it in a vehicle. For in its natural form it is a gas, which means that huge tanks are necessary, or at best large ones if high compression is used. Cryogenic temperatures are needed to liquefy it. It can be adsorbed in hydride form in porous metal tanks, but then a volume of metal "sponge" four feet square would be required to keep on hand enough fuel for some two hundred miles of travel.

Of all the various alternate forms, not a single one, with the exception of solar energy, holds the promise of being able to supply a significant amount of our energy needs in the near future. And even then it will be used, at least at first, only to provide hot water and space heat. Indeed, that is being done in many places already.[9] But for generating electricity, there are no really good ways of using the sun's rays as yet. Solar cells, which directly convert solar heat to electricity, are still far too expensive for domestic use.[10] Other techniques are still experimental.

Fusion power—the next step beyond present nuclear techniques—seems at least two decades away, and may never be possible. It has not even been proved that nuclear fusion, which is responsible for the sun's heat, can be harnessed on a controlled basis for use on earth. The use of antimatter and black holes are pure speculation at this time.

Energy to Burn

Suppose one or several such methods *are* perfected? What then? With enough energy, and enough technology, we could recycle every bit of material substance over and over again; we could not only remove every bit of pollution from manufacturing, vehicle, and power plant effluent, but we could even reuse the materials contained in the emissions; for a pollutant is actually a useful material in the wrong place.

Two nuclear scientists, W. C. Gough and B. J. Eastlund, have proposed a "fusion torch." Trash and garbage fed into such a device would quickly be turned into gas. There would be an automatic sorting out of all the basic elements, which would eliminate the complications of sorting a complex mixture of substances and compounds, such as plastics, metal alloys, wood, and so on.[11] This would be the ultimate in recycling, but it depends on fusion power being brought to fruition.

The only thing we could not take care of with unlimited energy is basic heat load, which brings us to what may well be the one real, physical limit on world growth. It should be clear by now that energy is everything. There is nothing we eat, drink, ride, wear, sit on, lift, sleep in or play with that does not require some expenditure of energy.

Use of energy always entails a conversion—from the chemical energy stored in wood or oil or coal to heat; from heat energy to mechanical energy; from mechanical energy to electrical energy; from electrical to mechanical or heat energy; and so on. But one of the basic laws of thermodynamics tells us that with each and every conversion, some of the energy is lost.

This loss of useful energy is really at the heart of what we call *efficiency*, which is defined as the amount of useful energy gotten out of a process, divided by the energy, or fuel energy equivalent, put in. A typical fossil-fuel power plant can convert not more than about 40 percent of the fuel it burns into electricity. The rest goes off as heat! Present-day nuclear power plants are only about 30 percent efficient; 70 percent of the energy contained in the nuclear fuel goes off as waste heat.

In each step of the total process—which includes extraction of fuel from the earth, transportation of the fuel to the generating plant, and transmission of electricity to the users—some energy is lost, with the result that the efficiency of the overall process drops to somewhere around 16 to 18 percent. A typical American automobile may deliver as little as 10 percent of the original energy contained in the fuel.

Oil heating of homes and offices is perhaps 60 to 65 percent efficient. The overall efficiency of energy use in the United States is just about 50 percent—which means that half of all the fuel we use is wasted. (How much of the rest is spent on unnecessary things is another question.)

The waste heat cannot simply be dumped into a nearby

lot, or even into the air. Vast quantities of cooling water are needed to handle it Already some 10 to 20 percent of the total fresh-water flow in the United States is used for power plant cooling. A large fossil fuel power plant (1 million kilowatt) needs 30 million gallons of cooling water per hour; an equivalent nuclear plant needs 50 million gallons per hour. Some of the heat from fossil fuel plants goes up the chimneys. Nuclear plants have no chimneys, and so all the waste heat must be handled by cooling water.

The waste heat warms the water by some 15 degrees F., and that warmed water goes into our rivers, lakes, bays, estuaries, and coastal waters, all of which are incubating and growth regions. But higher water temperatures can have devastating effects on water life. (There are other ways to handle this heat, such as huge water cooling towers, but they are expensive and have other effects on the environment.) It may, however, be possible to use some of this waste heat—for some kinds of fish farming, for warming soils in cold areas or for keeping northern waterways like the Saint Lawrence Seaway open all winter. If the plant is close enough to the user population, some of the excess steam can be piped to users for hot water and heat. But present laws tend to prevent such location because of the dangers and/or pollution.

The overall name for what is happening here is *thermal pollution*. This is the basic price we are paying for our electricity, indeed for our civilization.

One possible "fix" is to make a fundamental change in the way we produce our electricity. The present method involves a cumbersome, roundabout route that uses heat (produced by fossil or nuclear fuel) to convert water to steam, steam to drive a generator, and finally converts the mechanical energy of the generator into electricity.

But there are so-called direct conversion systems that are more efficient. We have already mentioned the fuel cell. Another direct conversion process goes by the name *magnetohydrodynamics* (MHD), and involves converting a moving stream of hot gases directly into electricity. The problem is that it must operate at some 5,000 degrees F., with effects on operating materials that can be imagined. Experimental devices have been built and operated, however, both here and in the Soviet Union.

There is the possibility of combining several processes, with a considerable overall increase in efficiency. Thus the hot gases that have already been used in the MHD generator can be used to drive a thermionic system and again in a thermoelectric device.[12]

Or, the MHD system can be used as a "topping" device for a conventional steam-turbine plant, meaning that its hot exhaust gases are used to drive the electric generator. Combining various processes in this way could result in increasing the efficiency of the generating process to perhaps 60 or even 70 percent.

But all of these methods are currently still in experimental or demonstration phases, or are being used operationally only in such applications as spacecraft, where economical operation is not the paramount factor.

It is important to understand too that *all* the energy we use is eventually degraded into heat; thus not only the waste heat but eventually all the energy generated must be dissipated into our environment. The importance of higher efficiency is that when less energy is wasted, less need be generated.

The major advantage to solar energy is that, in addition to its being free and available, it does not add to the heat load of the earth; it does, however, cause some redistribution of heat that could be troublesome.

At this point, man's activities still contribute only a small addition to the earth's total heat budget (solar energy in versus heat radiated out into space*). Currently the solar input is estimated to be some twenty thousand times man's contribution. It would therefore seem that there is really nothing to worry about in this respect.

There are two factors that must be considered, however. One is that while worldwide effects may not yet be a factor, local effects already are. Surface temperatures of urban areas are known to be several degrees higher than those of nearby rural areas. This "heat island" effect arises from blockage of air movement by tall buildings, absorption and storage of solar energy in streets and buildings, and of course from man's industrial, transportation, and other activities. In the Los Angeles area, the total amount of heat generated is about 3 percent of the incoming radiation, while in some highly industrialized areas of the northeastern United States, northwest Germany, and southern Belgium, the figure is already up to between 5 and 10 percent.

As urban islands become larger and perhaps merge, regional effects may begin to become an important factor. Effects on climate in these and nearby areas could begin to be severe.

After all, the solar input is, ultimately, the determining factor in our weather. Michael McCloskey, executive director of the Sierra Club, maintains that by the year 2000 energy produced by man may be up to 30 percent of the solar input, with the result that our population centers will turn into giant heat radiators and cause major disturbances in our local climates.

---

* Some heat from radioactivity taking place within the depths of earth must be added to the total heat load, but it is a small amount compared to the solar input.

The point is that we are becoming more and more capable—through sheer numbers, if nothing else—of unbalancing nature's systems. It may therefore become necessary to limit the population density and/or size of certain large urban areas because of their negative effects on climate and earth's ecosystems.

With better management, much of man's heat inputs could be routed to the oceans, which are large, cold, and, along the populated coastal regions, generally available. But it will be necessary to see whether large-scale dumping of heat into the oceans will have negative effects. Perhaps the water to be used can be drawn from great depths—regions that remain cold because they are unmixed by surface waves and therefore not heated by the sun. Then the water can be heated just to the temperature of the adjoining surface waters. The discharge would thus create no troublesome temperature differences. There is even a possibility that this would bring forth nutrients from the ocean deeps—would be a kind of artificial upwelling that, in its natural form, is helpful to ocean life.

Big Thinking

In 1967 *Fortune* Magazine reported that "scientists and technicians, in a frustrating search for some way to describe the changes that their work on nuclear energy portends, speak glowingly of air conditioning Africa and heating the subarctic." [13]

Most of us now shudder at the very thought; for both processes would release enormous quantities of heat into the ecosystem.

But for a long time this "big" thinking was typical of our approach to nature. Indeed, the more effect we had on the earth the prouder we seemed to feel. There was a kind of technological machismo at work, a feeling that every obstacle was a challenge to our "manhood" and had to be not only overcome but smashed.

The same has been true of our approach to resources. Our own generation is blamed for it, but it is a legacy left to us by our free-swinging forebears. There was a time when a frontiersman would kill a bison, cut out its tongue for supper and leave the carcass as a "leftover." Indeed, the American Indian used to stampede whole herds of buffalo over cliffs and then take the meat and hides they needed from a few.

Today we are faced with a shortage of desirable materials and a plethora of "leftovers," including heat. The question now is what to do about it.

First there is the technological approach. While we cannot recycle energy, as we can materials, perhaps we can make our industrial and other technological processes so efficient that our productive capacity can grow without a concomitant increase in waste or heat production. And perhaps if we learn to use one or more of the various other energy income sources (e.g., solar, wind, waves, tides),[14] we can even decrease the artificial heat load the earth is currently being subjected to, or spread it out so as to eliminate the heat island effect.

For a long time it has been "obvious" that energy use and economic growth are inextricably entwined. A major study, the Ford Foundation's Energy Policy Project, seems to have disentwined them, however. In a preliminary report,[15] it is suggested that a reduction in energy consumption can be accomplished without necessarily causing an era of unemployment and lower growth rates.

We may even help ourselves in the process. Think for example of the enormous waste of paper and energy involved in the payment of bills in this country. Billions of bills and checks are produced, and physically carried from place to place, each year. How much quicker and easier, and resource saving, it will be when we can pay by phone. Demonstration systems are operating in Connecticut and Minnesota. If they are successful we are likely to see a rapid growth of such systems, and a consequent saving of labor, paper, money, time, and, mainly, energy.

We must learn to use technology sensibly, and we must stop bulldozing our way through nature. Surely our options are not limited to the poor peasant who is buffeted by every kind of ill wind, or modern man, who will lop off the top of a mountain if it interferes with his view.

If we are guilty of anything, it is of practicing "careless technology." [16] We must change that to careful, or sensible, or intermediate, or friendly, or soft, technology.*

Soft Technology

John McDermott, writing in the *New York Review of Books,* put forth "Gresham's Law of Technology": "Within integrated technical systems," he wrote, "higher levels of technology drive out the lower, and the normal tendency is to integrate systems." [17]

But as in all laws that deal with people (and in essence this one does), there is always the possibility of repeal. The fact that the *Whole Earth Catalog* was such a success suggests at least an interest in "lower" technology. This book is not, as is often thought, antitechnology; rather it is a primer for people who fear technology. It is filled with technology, leavened somewhat with philosophical yeast. The objective is to show how technology can be used

---

* The term *soft technology* should not be confused with the word *software*, as used in the computer field. Software refers to the written programs, flow charts, and symbol systems, while hardware refers to the electronic and mechanical equipment itself.

in a friendly way. For example: "Coleman lamps are terrible—they hiss and clank and blind you, just like civilization. Aladdin is the answer if you need good light and 117 AC isn't around. . . (It does require some babying to keep the mantle from smoking up; it's like not burning toast.)" [18]

Philip Morrison suggests:

> There is a reaction occurring which is going to incorporate technology in a complicated and funny way into the popular mind. And this is all to the good; the more *Whole Earth Catalog*, the better. I do not like the people who go off into the woods and kill themselves because they don't know anything about nutrition, about bringing babies into the world, about fixing their wells. They think you can just do it by good will, by heart. That simply is not enough. [19]

The *Whole Earth Catalog* was the beginning of another way for the people born into civilization. Though still no more than a soft breath in a hurricane, a movement is afoot, a movement of people who are not just reading about this sort of thing, but are doing it. A few have actually gone back to living off the land, really roughing it. Most are trying to use technology in a sensible way. They have regained some respect for the land. All are conscious of the damage we are doing to the environment, and of the damage that that does, in turn, to us.

Today when a young graduate psychologist and his family chooses to live in an abandoned mining town, often without electricity or hot water, chopping wood for his fuel, and farming or doing some sort of craft work for a living, it is no longer strange or even threatening to many of us.

A friend of mine has been living this way on an island

in Maine for three years. He and his wife, both artists, love it. But the fuel problem almost did them in; he had to cut wood for *three hours every day* in order to have enough for the cold winters. This cut deeply into his time and energy, and seriously affected his work. After the novelty wore off—it *was* fun at first—he began to wonder if the advantages were worth the effort. Finally a friend bought them a gasoline powered saw. Now they can get all the wood they need in fifteen minutes.

Not everyone who tries this "new" way of life wants to give up all of civilization's pleasures; nor is it absolutely necessary. The objective is not to suffer, but rather to get away from the idea (if not the reality) of trying to dominate nature. We must, in other words, distinguish between soft technology and masochism. Perhaps the hallmark of the new way of life will be a windmill providing just enough electricity for a hi-fi set.

The beginnings of a turnaround are being seen in "civilized" areas as well. Once more architects are considering such items as shade trees, cross ventilation, prevailing breezes and overhangs in their designs, making possible the construction once again of homes and offices that do not require air conditioning from the first warm spring day until the next time the heat is turned back on.

At the University of Texas School of Architecture, a laboratory, the Laboratory for Maximum Potential Building Systems, has been set up. As they put it in a flier, "Ours is not a fear of technology, but a fear of the 'system.' We are concerned with providing systems alternatives through technology or technological applications on a small scale, relating to the masses (hopefully) on a practical basis, both financially and technically." Their emphasis is on

decentralist approaches toward future development

with self help as a major concern in the total building process. Proposed projects include: Reconstruction of a water-pumping wind mill, a savonius wind generator compressor, low-water-using bathroom, low-cost high-efficiency wood burning stoves, space heating tin-can solar collectors, tin-can solar hot water heater, 16-gallon-drum solar greenhouse, algae producing biogas plant [produces a burnable gas from organic wastes], solar still for fresh water production.[20]

Is it really necessary for the rest of us to send five gallons or more of laboriously purified water to perdition every time we void four ounces of urine or a couple of cubic inches of feces? Anyone who has had to use the typical "outhouse" says "Yes!" Rikard Lindstrom of Sweden says "No," and has actually come up with a reasonable alternative. Called the Clivus Multrum, it is a self-contained unit that handles (separately) toilet wastes and kitchen wastes, and converts them automatically into usable fertilizer. It saves ten thousand gallons of water per person per year. Not only that, but it saves problems with plumbing and piping. It has no moving parts, requires virtually no maintenance, and produces seventy pounds of fertilizer per year as a bonus. The heat caused by bacterial decomposition is used to vent all odors up through the roof, which is obviously an improvement on even the most modern water closet.

Several thousand units have already been installed in Norway and Sweden. Careful monitoring and testing show it to be perfectly safe, perhaps more so than water units in that disease bacteria do not leave the house. All in all it appears to be the first major breakthrough since the modern flush toilet was developed a century ago, an excellent example of biotechnology.[21]

The Clivus Multrum is by no means the first such device, however; Henry Moule, a British clergyman, invented an "earth closet" in 1860. Requiring some two pounds of dirt per person per day, it too was reputedly odorless and produced fertilizer as a by-product. It was manufactured for a while in the United States, but lost out to the convenience of the flush toilet.

What Price Convenience

Indeed, *convenience* seems to be the key word today— as in the modern term, *convenience foods,* meaning prepackaged and even precooked. Such foods are often less nutritious than fresh items, though they may be *more* nutritious than wilted fresh foods, and certainly safer than nonfresh meats. They certainly use more (nonhuman) energy, however. They must be carefully packaged and require chemical preservatives. Energy must be used to cool frozen foods artificially to a low temperature, and then often used again to reheat them. Many modern, frost-free freezers have *heaters* in them to melt the frost!

Yet when there has been a tossup between conservation and convenience, conservation has never even had a chance.

Has our definition of the word *civilized* become so hardened that it cannot be changed? Can we reintroduce the idea that a bit of inconvenience is a reasonable price to pay for insuring that we do not run out of resources, energy or heat sinks? It may mean doing away with energy-inefficient suburbs, and going back to the bifurcation of city living and country living.

For city living is, by the very nature of its compactness, far less wasteful than are suburbs (except perhaps for the few real giants). If the city is reasonably designed, people can live within walking distance of their work and recreation. At worst, there is a high enough population density to support public transportation. (Large multi-family dwellings are almost always less wasteful of energy than single-family units, if for no other reason than that there is less surface for heat energy to escape in the winter and enter in the summer.) If activities are centralized, even travel between city centers can be done by quick and convenient public transport.

The main impediment to positive change is positive feedback. The more we buy and use, the better, or so it has appeared. We are, as we have seen, growth oriented. Every time we buy something, the gross national product goes up, and we have been brought up to believe that that is good. What we need is some kind of negative feedback, meaning that the more we do to harm the environment, the more something happens to dampen this activity. That something must be direct, immediate, and in proportion to the harm done. In a city today, the poor person's eyes sting just as much as those of the rich person who may be ten times more responsible. There is no direct correlation.

Rationing Energy

To get around this, two Australians, W. E. Westman and R. M. Gifford, have proposed a rationing system based on allocation of environmental credits called *natural resource units* (NRUs) to individuals and organizations. By assigning a price in NRUs to every resource use and activity that causes environmental impact (including childbearing), "a new system of environmental trade-offs can be established—one which maintains maximum personal choice within overall environmental constraint." [22] By divorcing the allocation apparatus from our present money system, the possibility of rich people doing greater damage to the environment is eliminated. Their extra wealth can, the authors suggest, be put into increased services. This, interestingly, is exactly how the rich of old lived their good lives.

While the idea has some appeal, a number of objections have been raised: (1) it would be extremely cumbersome and probably expensive to maintain; (2) the allocations would be quite arbitrary, and if large numbers of people felt the allocations are unfair, the system might end up damaging the environmental movement n   than it helps.[23]

The first objection, I feel, is not significant. With increasing capability and decreasing size and cost of computers, it is likely that the system would not be terribly unwieldy, though it does indeed add another accounting system to the money system we now use. It also leads to total centralization, which some people are concerned about. It would seem that the second objection is the more telling one. How does one allocate the environmental impact of purchasing a car, taking an airplane trip, or having a child?

Perhaps a more direct route is to ration energy use directly. Since energy use is involved in every activity, and in every item, and since all forms of energy are physically interconvertible, there need be little arbitrariness in assigning impact values.

The amount of energy that goes into the manufacture of a car is known, and how much is required to clean up the mess produced thereby can be estimated fairly closely. (Some adjustment may be needed at the end of the year; or, alternatively, the "cost" of pollution control and clean-up can be taken "off the top," with only the remaining energy available allowed to be drawn on.) Gasoline and fuel oil rationing are automatically taken care of, as is electrical use, and so on. Water use is not as direct, but can be apportioned fairly closely.

Trips on public transport, except perhaps for aircraft, might be forgiven altogether, both to save the complication and expense of the bookkeeping, and to encourage their use.

Appliances that are energy conserving might be charged at a lower rate, and further savings would automatically show up in lower energy costs. The same holds for smaller cars.

Each person would be assigned a National Energy Ration Card at birth that, clearly, would start off with low rations in the early years, and increase as time went on to some maximum in the late teens. Everything one does or buys would be charged to the card. If you go overboard on food, you may have to use less for clothing and transportation. If you use wood fuel, which is replaceable, you may be charged less for each BTU. Families living together can apportion their common charges—for fuel, food, shelter—as they wish.

It is, admittedly, an enormously complicated, difficult, and all-pervasive system. But so is our money system. And if, as seems to be the case, voluntary approaches do not do much good, and since people with money are willing to pay any price for their needs and desires, this would seem to be the most equitable approach. Indeed, if abundant times ever really come (especially in monetary terms), we may be able to do away with the money system altogether, and use just the National Energy Ration Credit Card for our purchases.[24] If the allocations are low enough, this becomes a real possibility.

We will probably come to such a system only when we

are in dire straits. But if such a system were put into effect before that time, it would force people to think about their effect on the environment. It might even rekindle a revival of interest in technology—at least to the extent necessary for living sparingly.

Hunger and Malnutrition

For a while in the late 1960s it seemed that new agricultural techniques, going under the general name of the "Green Revolution," were going to push the plow ahead of the stork: newly developed seeds combined with intensive growing practices would produce extraordinary increases in yield.

The signs today are, unfortunately, less hopeful. Poor growing conditions, changing climatic conditions, and poor preservation and distribution techniques, coupled with rapid increase in world population, seem to wipe out whatever increases are accomplished in world food production. In 1974, world supplies of grain, a basic food in less-developed countries, fell to their lowest level in two decades.[25]

A report has been issued by the National Academy of Science, for example, warning that this century's warm climate may be an anomaly. There are indications that a swing to colder times has already begun.[26] A swing of only a few degrees would have enormous effects on world agriculture.

Technophobes can point to the negative aspects of the Green Revolution, especially the need for large quantities of chemical fertilizer and irrigation. Fertilizer production requires large quantities of energy, and the oil situation has driven up the price of fertilizer and created scarcities. There is also the problem of runoff; often some of the fertilizer runs off into the water supply, contaminating it.

Water for irrigation requires pumping (and the energy to do so), and such pumping may deplete a water supply that is not ample to begin with. When water is withdrawn from a subterranean supply, there may be problems with the land. Because the newer plant varieties have a higher moisture content, power is needed to dry them. And fuel and trucks are needed to transport both the chemicals and the crops.

If large quantities of a single crop are planted (the new techniques and seeds are mainly in the cereals and grains), they are open to large-scale infestations of pests and plant diseases, often requiring great quantities of pesticides. Often, too, only the big land owners can profit from such planting.

As a result, many writers have scorched the Green Revolution, calling it a sham and worse—totally ignoring the fact that it arose *because* of the fact that traditional agricultural practices were not doing the job. Mexico in the 1940s, for example, was a hungry land. Thus it is no coincidence that Dr. Norman E. Borlaug, who won a Nobel Peace Prize for his pioneering work in this field, should have begun his work in that place and time. The Green Revolution derives from the fact that a third of the world's population is estimated to be malnourished, while another third lives on the edge of starvation. As many as ten thousand people starve to death each week in the poor countries.[27]

The fact is that the Green Revolution gave the world some time. Had something been done about the growth of world population a decade or so ago when the Green Revolution began to have an impact, many areas that had become able to export food might not have fallen back again. But there are 80 million new mouths to feed each year! Further, as population grows, the less fertile lands must be brought into cultivation, requiring more energy and other inputs for the same output.

Population growth is not the only villain in this drama. Even if population holds still, we may still be in trouble.

People who are better off not only eat increasing amounts of food, but different kinds: more meat, for example. Unfortunately, meat animals, especially in industrialized countries, are fattened on grain and to some extent compete with humans for that grain. A pound of beef requires anywhere from five to ten pounds of grain, which means that all the food problems are multiplied several fold. This has been a significant factor in the present world food shortage.[28]

The meat situation does not have to work this way. Animals can be brought to maturity on forage, as they used to, rather than being put into feed lots by the tens of thousands to be fattened up. Animals can feed on grasses and other natural vegetation not edible by man, and so may *expand* the food supply.[29] But forage-fed cattle take a month longer to reach full size, and are not quite as desirable on the market.

It has been estimated that two-thirds of the global increase in grain consumption is due to population growth and the other third to growing affluence.[30]

Advanced Food Technology

There are a number of ways in which science and technology can help the less-developed countries without being destructive to the receiving culture. A great step in reducing malnutrition would be accomplished if scientists could achieve the same kind of breakthrough in the production of soybeans and other high-protein vegetables that has been accomplished with the grains (e.g., corn, rice, and wheat).

Another way is to develop species of grain with higher protein content. In some cases, especially rice and corn, this has already been done; unfortunately in most cases the taste, texture, or color vary somewhat from what the people are used to, so they prefer not to use them. Work along this line continues.

Although peas and beans do not grow as quickly as grains, they perform another trick that is most useful. They "make" their own nitrogen fertilizer; more specifically, bacteria in their roots enable them to draw what they need of this vital nutrient from abundant supplies in the air. If this technique could be transferred to the grains, it would save not only the petroleum products and the huge amounts of energy required to produce nitrogen fertilizer, but also the energy used in transporting it. There are indications that it can be done.[31]

We need improved weather forecasting, so that farmers can prepare for drought, heavy rains, cold, or heat.

We need to increase the efficiency of photosynthetic conversion of solar energy to plant food, presently only a fraction of 1 percent. One approach being worked on is to produce plants with leaf structures that will make more efficient use of sunlight and carbon dioxide.

And finally, we need to find better ways to combat deterioration, depletion, and erosion of the soil.

Pest Control

In pest control, we need to find ways to prevent the great losses of food to birds, insects, and microorganisms without using heavy doses of pesticides. Creating crops that are disease resistant is one way, though insects and disease organisms eventually manage to get around this.

The soft technology approach, for example, picking the bugs off by hand, or using the large number of other "home" remedies, sounds tempting, but will not work until and unless many more of us are willing to go back to being farmers. High yield requires practices that are either labor intensive or technology intensive. (Today less than 5 percent of Americans are farmers, whereas 50 to 80 percent of the people in other major food-producing countries work in agriculture.) Though fungicides, insecticides, and weed killers have their side effects, it has been estimated that as much as half the U.S. agricultural crop would be lost each year without them. You may have noticed what has been happening to food prices as it is.

Maybe the answer is less troublesome insecticides than the ones we have become used to. Researchers in Great Britain are testing a "clean" pesticide, one that is relatively free of the cumulative and toxic environmental effects associated with DDT and the other so-called chlorinated hydrocarbons. What is especially interesting is that the chemical mimics, but more intensively, the insect-killing properties that occur naturally in the pyrethrum family of plants (which includes the chrysanthemum).

Its promise lies in the fact that it is more active against insect pests than DDT, yet is nontoxic in animals and humans. Indeed, it is rapidly metabolized in the body. And the one problem that remains—that it is toxic to fish—may be gotten around by careful handling.

A newer technological approach, called *biological control*, uses viruses, hormones, radiation, or other biological methods to kill, sterilize, confuse, lure, or otherwise control the pest. Generally the idea is to use natural enemies of the pest to get rid of it, although synthetic hormones and radiation are being used in other interesting ways as well. Because the biological approach tends to be species specific (only the offending species is attacked), there is less general messing up of the environment.[32] In one approach special hormones are used to interfere with the insects' growth cycle, so that they die off without reproducing.

A special advantage of these methods is that the control mechanism simply disappears, or is biologically degraded, when the job is done.

Unfortunately this "breakthrough," along with so many of the others we keep hearing about, is having a hard time breaking through. One problem is expense. The testing for safety and efficacy, for example, is multiplied because, since each remedy acts on a single species, there must be many more remedies than when broad-spectrum pesticides are involved. Further, microbial agents cannot be patented as a pesticide can, because they occur naturally. Obviously, the markets for each are also limited. And, finally, there are fears that, if used on a wide scale, even safe items might mutate and pose a serious health problem.

In one of the more intriguing radiation approaches, large numbers of male insects are raised, sterilized, and released in the infected area. Since the females of certain insect species tend to mate only once, the objective is to get as many of them as possible to mate with a sterilized male. The technique has proved especially useful for isolated populations, as on the island of Curaçao, where in 1955 a test of the method eliminated screwworms. In the American Southwest, the approach has been used, though with less success because of continuing infestations from Mexico.

We in the United States can afford to denigrate chemical pesticides because insects are merely an inconvenience to us. But in other countries the mosquito, the tsetse fly, and the snail are matters of life and death. Malaria, transmitted by the mosquito, is probably the most widespread disease on earth. The World Health Organization estimates at least 100 million cases, and a million deaths from it, yearly. In other words, one of the world's major health problems remains to be solved.

And even in the United States we may be seeing a resurgence of problems. In certain areas of California, encephalitis-resistant mosquitos have become resistant as well to practically all pesticides used against them. And ravages to cattle in the American Southwest from the screwworm fly remains a very serious problem.

Dr. Frank E. Hanson, associate professor of biological sciences at the University of Maryland, Baltimore County, has been studying the feeding behavior of certain insects that feed on crops. It is generally thought that what insects eat is determined by genetics. Dr. Hanson's work suggests that insects can be "taught" to alter their diet, and eat the less valuable plants. The perfect weed killer.

Our Eating Habits

In our own country there is no real problem with food shortages; but there are still many undernourished and malnourished people, both youngsters and adults. Part of the problem is ignorance of what is a good diet. A small part is lack of funds to purchase a good diet. But the main problem is lack of interest. We seem unable to turn aside hot dogs and potato chips and soda and sugared cereals for more healthful if less attractive foods.

Perhaps we could find a way to handle the sugar molecule so that it acts on the tongue but then passes through the body without entering the tissues. There is hope that by making the molecule large enough, this can be accomplished. Or maybe we can develop candy, cake, and whipped toppings that are good for us.* This would certainly be easier than making tens of millions of people change their food habits. A new sweetener called *aspartame* has been developed that is a combination of two amino acids and is therefore technically a food.

Philip Morrison points out that there is one common element in industrial diets all over the world, from South Africa to Norway: food consisting of a starch within and a layer of crisp salted fat without. Our representatives are French fried potatos and potato chips. Prof. Morrison feels, as I do, that there is a lack of rationality in our food habits: "Everyone knows these foods are bad for you," he says. But he feels rationality will return because shortages will. I wonder.

In the meantime technology has managed to bring a wide variety of foods to our table. Frozen foods generally taste better and are more healthful than canned foods; more advanced technology may make an even more healthful diet possible for the many people who *want* the convenience, the time-saving and effort-saving characteristics of prepackaged, even precooked foods.

As for the "fact" that we are "filling our foods with poisons," again the truth seems to lie between the extreme positions taken by both sides. In general, when people die from what they eat, it's because of *Clostridium botulinium* or some other virulent organism that remains because of improper sterilization. And a fair number of those

---

* So-called health-food candy and cake, utilizing nuts, honey, sesame seeds, whole wheat flour, etc., are better but do not seem to have the wide appeal of "regular" desserts.

"poisons" put into our foods are there to combat spoilage —which includes the growth of these microorganisms.

Some preservatives are there for economic reasons. But in a certain sense, and paraphrasing that unfortunate remark of a General Motors executive, what's good for the distributors is good for us. One reason is that food which does not spoil quickly is less expensive to handle. Thus, while preservatives make life simpler for the people in the food business and save them money, they also save us time and money. And they make possible the sale of foods that would otherwise be out of season for a large part of the year, and maybe totally unavailable in certain parts of the country. (On the other hand it is not necessary for cereals to have a shelf life of eighteen months!)

But those who complain about preservatives and food coloring might consider using less of the convenience foods (especially superfluous items like candy, cake, ice cream, and soda!) and more fresh foods, or even raising their own.

The real question is how much of anything we eat. Too much of anything is bad for us. Dr. Sohan L. Manocha, Emory University nutritionist, says there are natural poisons in many of our most common foods, including potatoes (uncooked), cabbage, beans, fruits, fish, and meats.[33] Even that great all-around food peanuts contains a poison called aflatoxin under certain conditions.

And remember that chemical-free food is available to those who want it—and are willing to pay for it.

Birth Control

Technology by itself clearly cannot solve the population problem, but it can help by supplying a better contraceptive. There is none available now that does not have some drawback—unreliability, complicated use, medical problems, and so on.

The newest contraceptive method in wide use today is the intrauterine device, or IUD, which, with an effectiveness rate of 95 percent, has proved to be one of the most reliable. Its main advantage is that once it is inserted, there is nothing further the wearer must do except to examine herself once a month to make sure the device is still in place.

By contrast, the user of the "pill" must remember to take it for twenty-plus days each month. On the other hand, the contraceptive pill is the most reliable method of all, with an effectiveness rate of 98 to 99 percent.

The desirability of these two methods is indicated by their rapid acceptance. Although exact figures are not available, it is believed that about 4 million American women are using an IUD, and some 10 million are on the pill (50 million worldwide), though both approaches are relatively new.

This is in spite of the fact that both methods involve risk to the users' health! With the IUD there are instances of uterine perforation and infection, spontaneous abortion, as well as unknowing expulsion. As of June 1975, 313 septic abortions had been associated with use of the IUD, while forty-three deaths had been attributed to its use. This works out to a rate of 1 to 10 deaths per million users per year.

With oral contraceptives the death rate is higher, about 22 to 45 per million users per year. Hospitalization rates appear to be about the same for both methods: from 0.3 to 1.0 per 100 women users per year.

Problems with the pill include increased risk of stroke, blood clots lodging in the lung, and thrombophlebitis (inflammation of the veins). On May 3, 1975, evidence was presented in the *British Medical Journal* of increased risk of heart attack.

It must be emphasized that these are increased *risk* factors, not certainties of problems, and must be compared with (1) the possibility of pregnancy, which carries its own risks;[34] (2) other contraceptive methods, such as the diaphragm, condoms, and foams and jellies which require "getting ready" for the sex act, and which are in general less reliable than the pill and IUD; and (3) abstinence.

Clearly, convenient, reliable methods of contraception are important enough to millions of women that they use them in spite of increased risk to health. Clearly, too, we need methods as convenient and reliable, but safer. Other approaches, such as a morning-after pill or a simple, inexpensive, reversible sterilization procedure would be a great help. All of these are possibilities but need more work.

An interesting line of research is being pursued at the University of Missouri School of Medicine as a substitute for vasectomy, a form of permanent sterilization for men that has its own complications—hematoma, edema, infection, pain on ejaculation, sperm granulomata, vas recanalization, along with psychological side effects.

M. S. Fahim, D. G. Hall, and R. Der report: "In considering an approach to male contraception which would be neither surgical nor pharmacologic, the authors elected to reinvestigate, utilizing modern technology, the effect of heat in suppressing spermatogenesis without alteration of Leydig cell secretion." [35]

Using various methods, such as warm water, infrared heat, and microwave diathermy, they warmed the testes of male rats for various periods of up to fifteen minutes, and were rewarded with as much as seventy-five days of sterility.

They point out that "Histological studies of the testes revealed an absence of spermatogenesis. No significant differences in blood testosterone levels occurred in treated animals as compared with controls." With ultrasound as the energy source, they found:

> The treated animals have not impregnated females as of the time of this report (six months following treatment). No side effects have been noticed. The histological studies were identical to those produced by microwave. No hormonal imbalance and no change in male behavior or libido were noted.
>
> Longterm affects of recurrent heat application by modified electronic equipment are currently under investigation.[36]

People will only use birth control if they want to. And one of the major truths is that people, all over the world, *want* more than two children. But we have also seen that the birthrate seems to drop almost automatically as a country begins to move from the status of less-developed to developed country. It would seem then that industrialization will (eventually) play a big part in controlling world population.

Fire!

Every minute there is a fire in an American home; every day there are 170 deaths or disfigurements from this scourge; our loss of life from fire is two and a half times that of the United Kingdom and six times that of Japan!

Smoke and toxic gases cause many of those deaths and injuries. In experiments at the National Bureau of Standards, fifteen different floor-covering materials, including nylon, acrylics, polypropylene, vinyl, wool, and oak (both varnished and unvarnished) were tested. All of them gave off dense smoke and varying amounts of carbon monoxide, a poison, and the flammable gases acetylene, ethylene, and methane.

Do we really have too much technology? Can we not develop and force the use of fire-resistant materials that do not produce toxic gases, and that are self-extinguishing?

The textile and apparel industries, which had until recently opposed either mandatory or voluntary rules regarding flame-resistance standards for their products, are now showing greater interest in such regulations. The major problem has been cost, particularly in the highly competitive children's line. Walter A. Haas, chairman of Levi Strauss, the world's largest maker of apparel, estimates that the company's line of boy's jeans and tops would cost 20 to 30 percent more in flame-resistant material, but reports that the firm has decided to introduce a line of flame-resistant boy's pants nevertheless. The problem of course is that past market research indicates that the public will not be willing to pay the higher price. Perhaps greater consumer awareness will change this attitude.

Further research may help by cutting down the differential. Researchers at the U.S. Department of Agriculture have come up with a flameproofing process that appears to be both effective and economical, though at the moment only for wool, wool-blended fabrics, and nylon. The new treatment could help eliminate some of the three to five thousand fatalities caused every year by flammable fabrics, and such a treatment could be used not only for clothing, but for blankets, carpets, and airline upholstery—all applications for which flame resistance is truly a matter of life and death. Most encouraging of all, it has been estimated that only ten cents' worth of the chemical used (tetrabromophthalic anhydride) is needed to treat an ordinary skirt! [37] As an added bonus, tests showed that treated wools were more resistant to moths than untreated ones.

Health Care

With the use of television links, data- and voice-transmission techniques, and the computer, health care in sparsely populated sections of the country can be vastly improved. For example, with a visual communications system, a physician could supervise treatment being given a patient far from the physician's office.

Touch sensors can be added, along with specially built teleoperators;[38] not only would the physician be able to see the patient, and listen to his heart and lungs, but he could even feel for swellings and perform other diagnostic techniques dependent on the sense of touch.

But it has been pointed out, and it is quite true, that better environmental control and personal action may be even more important for health care than improved or increased biomedical techniques.

A student team made up of seniors from the University of New Mexico medical school went to work in a group of clinics set up in isolated areas that originally had had no medical services. For three-month periods, the students lived in the assigned area and gained experience as well as satisfaction. They consulted by long-distance phone with more experienced personnel "back home." Would the program have worked without this contact? The students might have done more harm than good.[39]

Most of the advanced technology created for health care is designed for use in large medical centers, not in decentralized medicine. We may need to develop methods and equipment that are more suitable to rural practice: perhaps just simplifying certain of the methods now used in the large centers would be a good first step.

We have heard much about transplants; these are exciting, and have received much publicity in the news media. Less widely publicized are advances in diagnosis and treatment that may make such drastic steps unnecessary. Prevention is the direction that medicine of the future must take.[40]

But even with respect to today's techniques, there is a widespread feeling that our medical system in general is not delivering the health care we desire. The fee system we are used to tends to divide the populace into two fairly exclusive categories—well and sick. It is not until we are sick, generally, that we go to the doctor. But a growing emphasis on insurance and prepayment plans, and perhaps even national health insurance, promises a demand for care that is likely to mean that the truly sick will have greater difficulty than ever in getting treated. For then not only the sick but the worried well and the getting-sick will be involved, not to speak of the well who want to stay that way.

Dr. Sidney R. Garfield, of the Kaiser-Permanente Medical Care Program in Oakland, California, says that "With this altered demand, instead of caring for the sick, the doctors are spending a large portion of their time trying to find something wrong with well people, and they are doing this with technics [sic] taught them to diagnose sickness." [41]

He points out that the only way to medical care today is through the physician, which he maintains is not only wasteful of medical manpower but a serious bottleneck as well. At Kaiser-Permanente (a prepaid medical plan), a new system is being developed, the heart of which they call *health testing;* it combines a detailed automated medical history with comprehensive panels of physiological and laboratory tests administered by paramedical personnel, physical examination by a nurse, computer processing of results, and a doctor review.

It may be that only with increased use of technology can the goal of improved personalized medical care on the scale envisioned be accomplished.

Experimental Cities

It is easy to accept the usual beliefs about modern life—that industrialization, affluence, and particularly city life are at the heart of today's social problems. But there is no real evidence that urban dwellers suffer any more from mental problems than do those in the more rural areas. Similarly, the divorce rate is highest among the *least* affluent Americans;[42] we can therefore expect it to decrease, all other things being equal, with rising levels of income and education (which are usually higher in rural settings).

And, finally, it is not at all clear that present-day crime is in any way directly connected with industrialization or even large size of cities. It is well to remember that one of the main advantages of early cities was mutual protection against marauding bands of robbers and murderers. In the nineteenth century travelers between Newark and New York were being waylaid in the cedar forests of New Jersey through which they had to pass. When all else failed, the forests were burned to clear the area. Consider too that New York, the largest city in the United States, has by no means the worst crime rate, and that Tokyo and Hong Kong, among the largest and most densely populated cities in the world, have very low crime rates. It would seem there is good reason to doubt a direct cause and effect connection.

I would also maintain that while the city surely creates some problem people, it also acts as a relief valve for others, is indeed a haven for many who cannot fit elsewhere. Just as swamps are natural filter systems for poisons,[43] so are cities filters for social poisons, and are therefore abhorred for many of the same reasons that swamps are.

But how do we keep American cities from being noisy, congested, and crime-ridden? Does the answer lie with the "new town" concept? This involves planning, financing, and building a town to a given size and design—including population, the proportions and total area of the town, and even the areas to be given over to industrial, commercial, residential, and public facilities and to open space.

Great Britain and the Scandinavian countries have been particularly successful in this; England has built some two dozen new towns since World War II. But they are mostly still quite small, generally no more than a few tens of thousands of people. This approach cannot handle the vast numbers of people we in the United States are talking about unless great numbers of these towns are built. American planners also tend to differ with the European ones in feeling that larger entities are necessary to provide the cultural amenities and the wide choice of jobs urban dwellers have come to expect.

In the 1960s brave plans were made for the United States along these lines. One report called for the establishment by the end of the century of a hundred new towns, averaging a hundred thousand inhabitants apiece, and ten cities of a million each.

Toward the end of the sixties, it really looked like the new towns/new cities movement had taken hold as a national venture. Congress passed the housing acts of 1968 and 1970, which offered a well-leavened mix of aid for private new town developers. A new mood was evident in America, one concerned with the style and shape of the country.

Yet federal support has been limited to the issuance of federally guaranteed bonds; there are presently only sixteen new towns participating in the federal program, and a number of these are in serious trouble. There are some privately funded new towns, such as Reston, Virginia, Columbia, Maryland, and Lake Havasu City, Arizona, which are still moving along well, however.

At the 1975 meeting of the American Association for the Advancement of Science, a group of men interested in the new town movement got together to talk and lecture about "Innovation in New Communities." The general consensus was that there is little hope for any real movement and innovation unless and until the federal administration exerts some strong leadership. "Maybe," said Lou Manilow, manager of Park Forest South, a new community of some six thousand people outside Chicago, "the virtue of the sixties was in showing us that our optimism was absurd."

Perhaps. And the truth is that we still know very little in the area. Edward J. Logue, urban development expert, pointed out for example that the idea of a planned community is by no means new in the United States. Philadelphia, New Haven, and Savannah all started out as planned communities. On the other hand, San Francisco and Boston—said to be two of the most livable cities in the United States—did not.

Perhaps new communities smack too much of rationalism, of planning, of technology. But what is the alternative? The sprawl we have become used to? One estimate has it that a carefully planned, concentrated new town can save 50 percent in transportation, energy use, and various other requirements over the usual suburban community.

Mr. Logue also pointed out that continued suburban sub-division will result in continued racial segregation, with more and more explosive effects. He believes it is essential to keep this from happening.

An important aspect of new towns is the fact of voluntarism—you choose to live there, so you also choose to participate in the way of life it offers, which by the way tends to be a participative one. At the very least new towns offer an alternative to our present way of life. At the most they can be experiments in living, new technology at its very highest and most socially oriented.

For example putting in mass transportation is a social issue; buying a car is a personal one. But choosing to live in a place that forbids cars is also a personal decision. Yet how many places are there that provide this choice? Should there not be towns and cities that offer such an option? Does not this still rich and powerful country have the capability to do so? *

New communities offer ways of experimenting with, and living with, large-scale cooperative energy schemes, perhaps based on solar and wind technologies.

Alternatively, we may find that a new push in the direction of an alternative energy source such as oil shale will call for the creation of communities for thousands of workers and their families in new sections of the country. Shall we use this as an opportunity to create viable towns and cities, or shall we simply permit the same sprawl that arose when cities grew out of old logging, mining, or other industrial centers?

In Japan, for instance, a new science city, Tsukuba New-town is being built north of Tokyo. A number of research laboratories are being created and/or transferred there,

* Roosevelt Island, an apartment development in Manhattan's East River, is a move in this direction.

and a large university is being built. From an original population of about eighty thousand the final figure is expected to expand to about two hundred thousand. Of these, six thousand will be scientists, engineers, and educators. Tsukuba will be an experiment in several ways. Some fifty thousand of its population will be housed in a planned high-density, high-rise complex that will include not only apartment dwellings but downtown shops and entertainment areas as well. The university too will try something new. One-fifth of the students will be enrolled on the basis of high school performance rather than on the traditional strenuous entrance examination procedure. The performance of the test group will be compared with that of the other students; if they perform satisfactorily the method will be expanded to other universities.

Demonstrations of new ways of waste management, communications, building, health care, perhaps cooperative practices of various kinds, and many other areas of interest become realistic possibilities in new communities. And lessons learned in them could easily provide useful information for existing communities.

The problem with just about all of the present new communities in the United States is that they are small, and/or they depend upon some nearby larger city for much of their culture, jobs, and services, thus making them recipients of the cities' emotional and psychological problems as well. They are in general quite conventional, in most cases really only glorified housing devolopments.

Suppose the government put its resources behind a new *city*, including tax advantages to entice new industry, housing subsidies to attract lower-income groups, education subsidies to provide education and training where needed, and maybe even tax advantages to inhabitants for the first several years (to be sure of getting a well-balanced mix). It would be very expensive, but not nearly as expensive as welfare, crime, and poor health are today. And if it worked, private industry would be more likely to be interested in building other than conventional bedroom communities for nearby cities.*

There is, after all, a need to know just how pleasant life can be in a real city, to see whether city life can be designed in such a way, and whether modern technology can be applied in such a way, that the quality of life, even if no one can define it, is recognizably high.

Advocates of two experimental cities—one in Minnesota and one in California—have been hoping to design a "free-standing" city from the ground up, to apply as much modern technology and as much modern knowledge about man and his characteristics as possible, and to have a true mix of people. At the moment both lie somewhere between a hope and a dream.

The California project, Experimental City I, is conceived to be a model community/university of 30,000 to 50,000 people committed to testing theories of a more humane social and physical environment. It is expected that all property will be held in common, for example.

The Minnesota Experimental City, or MCX, is a larger concept, originally projecting a maximum population of 250,000 by 1985. Its backers have been pushing the project for about six years, but various barriers have arisen and stopped all progress. A group of less than a thousand persons presently living in the 78,000 acres of Minnesota wilderness picked as the site have fought the project. These people, some of whom moved to the area to get away from the city, want the area to remain what it is

---

* Though places like Reston, Virginia, do have industry, they offer little for low-income groups. Industrial workers tend to commute in, and the better-off to commute out, to nearby cities.

—pristine, rural, sparsely populated. Their resistance, combined with the downturn in the economy and an apparent weakened national commitment to the whole idea, puts the fate of the MCX very much in doubt.

To my mind the building of these, or any, experimental cities could be one of the most important tasks we could perform in the balance of this century. For we have to contend, not only with population growth, but with a great and constant wave of emigration from rural to urban areas as well.[44] If the latter trend continues, by the turn of the century two-thirds to four-fifths of all Americans will be crowded into just twelve urban centers (megalopolises) on less than 10 percent of our total land area. Over half the people—perhaps 140 million—will be jammed into the three largest regions: Boston–Washington, Chicago–Pittsburgh and San Francisco–San Diego. One estimate suggests that by the turn of the century as much urban building will have to be carried out as was done in all of man's history to date.

Clearly, if present cities and their surroundings take the major brunt, conditions can only deteriorate; if existing towns and small cities simply grow, as they have in the past—unplanned, haphazard—they will merely be repeating the mistakes of the past.

It seems far more desirable to see whether we have learned anything in our several millennia of city history; and whether we can apply what we have learned. The only way would seem to be to start from the beginning and plan out the whole thing. How much planning can be done before the results are truly stifling (to Americans)? We really don't know. Maybe a working, independent, full-sized city really can't be planned and built from scratch. But it would seem important to try to find out.

Rural Areas

Another way to prevent the total inundation of our urban environments would be to make our rural areas more desirable. A recent Gallup poll indicates that half of all Americans would like to live in a rural environment, though only about a third do at this time. The main reason for the imbalance seems to be an economic one—the rural areas, particularly the agricultural areas, cannot provide the jobs and promotion opportunities that can be found in nearby cities. Additional problems are lack of health care, and cultural deprivation.

Dr. Peter C. Goldmark, inventor of the long-playing record, the first practical color television set, and the video tape cassette, believes in rural society and has turned his inventive genius to the rural problems. It is his belief that communications technology provides the possibility of turning the whole thing around, of making rural areas highly desirable places to live. He contends it is possible to supply electronically everything that one can get in an urban environment; this, added to the advantages that rural areas do offer, could stem the tide.

To this end he is directing a project, known as the New Rural Society, which is receiving some small-scale testing in a few towns in a northeastern Connecticut planning region. There are four basic areas of concern:

1. *Education.* He suggests that teaching with the aid of satellite technology can convert the little red schoolhouse into the kind of educational experience that only metropolitan centers offer today, and can bring the classroom of a large university to interested students in outlying areas. Many colleges are offering credit already for television courses and programs.

At present levels of commercial technology, question-and-answer periods can be arranged by phone, or classes can be gotten together in regional locations one or more times per term. Exams can similarly be given in local groups, or by mail or even as part of the television programming, to be mailed in.

With two-way, interactive television—and cable TV makes that a realistic possibility—there can even be effective channels of communication from public to government, of which there is a remarkable lack at present.

2. *Culture.* Doctor Goldmark points out that all of the cultural events—opera, ballet, plays, concerts, even museum visits—can be relayed via satellite and cable television to people who are interested.

3. *Jobs.* We seem to be moving toward, or have already become, a postindustrial society: the basic center of our economy is moving more and more away from manufacturing, and toward such services as insurance, banking, and education, all of which can be handled electronically! [45]

If, in trying to make production work more interesting, we move away from the mass production techniques and toward small group assembly, some decentralization —smaller plants in outlying areas—becomes a distinct possibility. Nor need the business conference require that the management people all be conveniently near each other all the time, or even that the plant be near a city or airport. The conference can be handled by video techniques already in existence. A Connecticut bank has actually set up such an audio conference room. And government officials, businessmen, officers of the Salvation Army, a librarians' association, and a community service group have successfully carried out regular business meetings by "teleconference."

Dr. J. R. Pierce of Bell Telephone Labs has even suggested that in the future it will not be at all unusual for us to "communicate to work." The significance of this is seen in Dr. Goldmark's point that half of all gasoline use goes into daily commuting between job and home.[46] The New Rural Society, with its job-near-home, or job-*within*-home, could save much of this.

4. *Health care.* As we mentioned earlier, with the help of modern communications techniques, doctor's care can be brought to the isolated patient.

The Environment

Sir Peter Medawar, Nobel laureate in medicine and physiology, has stated:

> The word *ecology* has its root in the Greek word *oikos* meaning "house" or "home." Our future success depends upon the recognition that household management in this wider sense is the most backward branch of technology and therefore the one most urgently in need of development. An entirely new technology is required, one founded on ecology in much the same way as medicine is founded on physiology. If this new technology is accepted, I shall be completely confident of our ability to put and keep our house in order."[47]

But in order to do this we must have better data; that is, we have to know more precisely than we do now just how severe our problems are, so that we may better know where to direct our efforts. Betsy Ancker-Johnson, Assistant Secretary of Commerce for Science and Technology, suggests, for example, that our emission control problems may be somewhat exaggerated, mainly because emission requirements, especially at lower concentrations, are not very accurate. To be on the safe side, we call for un- necessarily stringent, and expensive, controls. We need instead, she says, better ways of making these measure- ments.

Our efficiencies in all kinds of processes can be raised, and sometimes we can combine two or more processes and thereby aid them both. It takes energy to produce food, and it takes energy to dispose of municipal wastes. Experiments have shown some remarkable ways to save energy on both ends. At the Oceanographic Institution in Woods Hole, Massachusetts, algae grown on municipal sewage are being fed to shellfish. In the process the sewage ends up cleaner and the shellfish end up bigger. Penn State has been spraying treated sewage water over crop and forest land for several years and achieving increases in crop production without harmful effects.

The high cost of oil has spurred development of methods of recovering energy from solid wastes. The Council on Environmental Quality reports that three cities have such processing facilities under construction and eighteen more are considering them.

Chicken and steer manure are being converted into edible animal protein. It seems nature's processes are not always perfectly efficient, and digestive systems leave quite a bit of protein in the food passing through—enough so that processed chicken and steer manure can augment animal feed supplies.

At the Canadian Combustion Research Laboratory in Ottowa, a mechanically simple device called an *air swirl generator* is being developed that promises to provide a blue furnace flame rather than the usual yellow one. The result, hopefully, will be significant fuel savings and less pollution. Estimated to cost about $1.50, the device could provide a possible saving of 10 percent in the user's heating bill. Car owners too may benefit.[48]

If other state governments won't emulate Oregon's bottle return law (forbidding no-return, no-deposit bottles), we can at least develop bottles that degrade when left outside. There are experiments along these lines, and with "nutritious" plastic packaging as well (nutritious to soil, that is).

Perhaps we can also do something about noise. The owl is the quietest bird in flight; his quietness enables him to "sneak" up on his prey. Do aircraft have to be as noisy as they are? Experiments being conducted on rotors with serrated edges somewhat similar to those on the wings of the owl tell us, "Maybe not." Prompted by some NASA research, the General Electric Company has begun a study of the phenomenon for possible applications in jet engine design.

We need "humanized machines." Wes Thomas, editor of the future-oriented newsletter *Synergy Access* and a computer expert, once told me of a "dream" he has. "What I'd like to do," he said, "is develop computer systems that are more 'human,' that people are not afraid of, in fact that they would get along with and even enjoy." He would like, therefore, to develop the "Fuzzy-Duzzy."

Today, he explains, computer terminals are made out of metal and plastic. They are cold and uninviting; therefore most people are immediately turned off by them. And they look menacing, like something a mad scientist would create.*

With Fuzzy-Duzzy, he says, "you would put your hands into this warm, inviting Teddy-Bear thing, and you would

---

* A computer-leasing firm has already begun to "humanize" its machines, however. The computer equipment can now be supplied already decorated with a choice of paintings. Everything from a tennis scene for the sports minded to an abstract design is available.

be able to look inside through a porthole. Inside there would be these big knobs you could get hold of and turn, instead of the usual miniature keys everybody keeps making mistakes on. By moving things around, you'd be able to communicate with the machine.

"And the pictures that come on the screen would not be the usual angular shapes, but nice, round organic forms.

"So I'm interested in developing a sort of organic computer terminal that people will feel at home with immediately." [49]

When I mentioned this to an acquaintance, he became furious. He called it underhanded, meretricious, and worse. A machine has no right being "friendly."

Thomas's "dream" is a bit extreme, I will admit. Yet humanized machines combined with people who are not afraid of them (and who understand where and how to use them) may be the way we will eventually do many of the things that remain to be done in basic education and job training, in health care and perhaps in many other applications as well. Just as the supermarket put the customer to work, so too may it be necessary for patients and prospective patients to do some of the pre-entry work themselves, aided by computers—humanized ones, of course.

Machines will also help us provide sight for the blind and hearing for the deaf, mobility for the lame and dexterity for the handicapped. Will humanization of such machines be necessary too?

We need, clearly, far more knowledge about ourselves —physically, emotionally, socially, mentally. We need a biology of the mind. We need to know more about such processes as meditation, altered states of awareness, psychosurgery.

René Dubos points out that "we know almost nothing of the processes through which every man converts his innate potentialities into his individuality. Yet without this knowledge social and technological innovations are not likely to serve worthwhile human ends." [50]

There is so much yet to be learned, to be understood. There is so much yet to be *done*.

11 *Fear and the Future*

He that will not apply new remedies must expect new evils;
for time is the greatest innovator.

> —*Francis Bacon, "Of Innovations"*

Loren Eiseley, author and anthropologist, once called
man a planetary disease.

If the point is that man's activities are harmful to the
environment, then it is not just billboards and highways
and skyscrapers that are harmful; everything we do becomes part of the disease. For farming is not natural;
parks are not natural; roads and lodges on the edge of
Grand Canyon are not natural.

Eiseley is a talented, creative, thoughtful person who
loves the land, and it is easy to sympathize with his feelings. Yet one wonders whether, if he had the choice, he
would do away with *all* of mankind.

That would be the only sure way of curing the disease.

There are alternatives, but all have serious drawbacks.
Doing away, for example, with those people who are dirty
and unproductive, or those who litter and keep their
thermostats at 74 degrees, or even merely those who are

concerned with only their own comfort and safety—which means the great mass of humanity—seems rather harsh.

Some sort of amelioration of the planetary condition would obviously be accomplished with the elimination of all "bad" technology and the retaining of the "good." But since there is very little agreement on what is bad and what is good technology, this approach is obviously impractical, at least for now.

A third alternative would be simply to wipe out all technology, which is what some technophobes have suggested. In general it is a case of their not having the slightest notion of what this portends; it is the simple answer. But as Emerson once pointed out, "Be careful of what you want. You may get it."

Suppose, then, the dyed-in-the-wool technophobes got their wish. Norman Cousins postulated such an outcome just about thirty years ago. Although he was writing with the specter of the just-dropped atom bomb in mind, and in support of world government, his words ring just as true now with respect to the negative effects of technology on the environment.

His approach, he wrote,

> is fairly simple. It requires that man eliminate the source of the trouble. Let him dissociate himself, carefully and completely, from civilization and all its works. Let him systematically abolish science and the tools of science. Let him destroy all machines and the knowledge which can build or operate those machines. Let him raze his cities, smash his laboratories, dismantle his factories, tear down his universities and schools, burn his libraries, rip apart his art. Let him murder his scientists, his lawmakers, his statesmen, his doctors, his teachers, his mechanics,

his merchants, and anyone who has anything to do with the machinery of knowledge or progress. Let him punish literacy by death. Let him eradicate nations and set up the tribe as sovereign. Let him, in short, revert to his condition in society in 10,000 B.C. Thus emancipated from science, from progress, from government, from knowledge, from thought, he can be reasonably certain of prolonging his existence on this planet.

This can be a way out—if "modern" man is looking for a way out from the modern world.[1]

When put in these terms, of course, doing away with technology seems a little less desirable as the road to utopia.

Besides, with all the complaining about what technology has done to us, we are still far, far better off than the countries that lack it. Wondering whether they, or we, would have been better off had the Industrial Revolution never begun is pointless speculation. It happened. As a matter of fact just as mercantilism was a logical outgrowth of feudalism, and capitalism was a logical outgrowth of mercantilism (indeed one of the objectives of capitalism was to cut down on cutthroat competition!), so too was industrialism a logical outgrowth of earlier technological development.

It can be considered an experiment, a tentative probe of society into a new area. Industrialism is only two hundred years old, and, if it can clearly be shown that we would be better off without it, there is at least the possibility that in future generations it can be done away with, or moderated.

On the other hand, Paul Goodman once pointed out that the scientific way of life, like communism or Christianity, has never been tried, that we are living with the horrors of a good idea badly executed.[2]

If we have lived too high, it has been out of ignorance, not maliciousness. Perhaps rather than castigating ourselves, we should simply be thankful for having been around at the right time. It was fun while it lasted.

What counts is not so much what we have done up to now, but rather what we do from now on. Fortunately a major change seems to be taking place in American life, though as with many major changes, it is starting small. In the face of an economy that has been stressing competition, big cars and cosmetics, we see a burgeoning demand for change, for a better quality of life. What we are after is a move in the right direction—one that will enable us, for the first time in history, to combine freedom from want with freedom to do one's own thing.

How and when do we get started? I believe the current turmoil is fair evidence that we have already begun. In our own country much of it comes out of earlier thoughtlessness, repression and unrestrained viciousness. Today these things still exist but are not national philosophy and are lessening. Worldwide, much of the turmoil is traceable to the fact that peoples everywhere are struggling out of bondage to freedom. Chaos is inevitable in a movement toward a civilization that works for everyone. It is an age of adolescence. Let us not give up on it just yet.

The point is not to put the blame in the wrong place, and kill off what may be the only controllable element, namely technology, in the process. Technology, like science, is boundary free and can help cement national friendships, which are sorely needed.* Few countries are interested in our philosophy or religious aspirations. But our radios, television sets, and cars, not to speak of our

* The recent Soviet/Amercian hookup in space is a good example.

agricultural and industrial productive abilities, are interesting indeed.

Consider too that many of the less developed countries are using the Western world as a model for their own development. If for no other reason, it is incumbent upon us to find a way to combine industrialization with humanity.

A step along the way might be to stop getting so much of our new technology from the military. We must understand the difference between tools and weapons. We should accept, indeed demand, the first; reject the second.

We tamed a wilderness in search of a better life, and found it wasn't enough. We built the greatest productive establishment in the history of the world, and are finding that this too is not enough. Now we are trying a new approach. We are moving from design specification to performance specification. We no longer say to government and industry, "Give us a steak in every microwave oven, and a Cadillac in every garage, and we'll be satisfied." For we have seen that that did not work. Now we say, "There must be more to life. Do something about it!"

Clearly, what we need is some new input on how to cope with the realities of life. And one of these realities is the continuing importance of technology in modern society.

We need to know more, then, not less, about technology. It must be introduced as a subject into our schools; it is not less important than government, language, or art. Making believe we are genteel and humanistic does not help. Refusing to get our hands dirty does not help.

If we must have a simplified model of technology, let us use the well-worn analogy of technology as a servant, but keeping in mind that it helps eliminate the degradation and humiliation that servitude usually involves.

Is technology a workhorse, or a wooden horse? A servant or a monster? Obviously it has elements of all these. But let us not try to keep this monster/servant locked up in the closet when not in use. It's never not in use. We must think about it: What is a good machine? Is there such a thing? Can a machine be used as a model for human life and behavior, and so help us understand ourselves better? How about in the future, when machines will become even more complex than they are now?

Let us not resist developing new, needed technologies for fear of their going out of control. With all the warnings about the railroad, no one smothered when the trains reached thirty miles an hour, or sixty miles an hour. And today this prototypical machine, this huge, snorting monster, is being touted as the answer to our transportation problem!

Let us set goals. Let us look for a new vision of humanity. Then let us use as much of technology as is necessary to take us there. If our problems are greater than they have ever been, so are our resources.

Those who wish to indulge in nostalgia and pessimistic prophesy should certainly be permitted to do so. But let them not be our prophets for the new age.

NOTES

Preface

1. C. P. Snow called attention to the problem in his book, *The Two Cultures and the Scientific Revolution* (Cambridge: Cambridge University Press, 1961).
2. See "Science IQ Dropping Fast," in *Science Digest*, June 1975, pp. 18–20, and "Knowledge of Science Declines," *Science News*, 29 March 1975, p. 206.

Chapter 1—Future Shock

1. In *Technology, Science and History* (London: William Heinneman, 1972), p. 123.
2. In *The Coming of the Golden Age: A View of the End of Progress.* See also Bentley Glass, "Science: Endless Horizon or Golden Age?" *Science*, 2 October 1970.
3. *Future Shock* (New York: Random House, 1970), p. 54.
4. Ibid., p. 73.
5. *Bulletin of the Atomic Scientists*, March 1975. p. 27.
6. Ibid, p. 23.
7. *Future Shock*, p. 22.

8. In *Science and Industry in the Nineteenth Century* (Bloomington: University of Indiana Press, 1970), p. 176.
9. *The Machine in the Garden* (New York: Oxford University Press, 1964), pp. 31, 32.
10. In "Garden of Eden—Updated," Leisure and Arts Section, *New York Times*, 24 March 1974, p. 24.
11. *Science and Industry in the Nineteenth Century*, p. 153.
12. *Science and the Modern World*, p. 91 (paperback).
13. *Harper's Weekly*, 28 February 1903.
14. New York: Putnam, 1966, p. 29.
15. *Science News*, 19 January 1974, p. 35.
16. From Bell's wide-ranging book, *The Coming of Post-Industrial Society, A Venture in Social Forecasting* (New York: Basic Books, 1973).
17. In his book *Industrial Research and Technological Innovation.* (New York: W. W. Norton & Co., 1968).
18. Reported in *Science News*, 14 April 1973, p. 238. See also Joseph P. Martino's articles, "The Pace of Technological Change," and "Adopting New Ideas," in *The Futurist* (April 1972 and April 1974).
19. Martino, "The Pace of Technological Change," p. 71.
20. Private interview, October 17, 1973.
21. Quoted in Dr. Dubos's column in *American Scholar*, September 1973.
22. Augustus Jessopp, *Arcady, For Better or Worse* (London: T. F. Unwin, 1892), p. 193.
23. Ibid.
24. The Poet Laureate of England, about 1830, ibid., p. 169.
25. Thomas Carlyle, "Signs of the Times," 1829.
26. Quoted in Marx, *The Machine in the Garden*, p. 214.
27. "Signs of the Times."
28. Quoted in Edwin Emerson, Jr., *A History of the Nineteenth Century, Year by Year*, vol. 2 (New York: P. F. Collier and Son, 1902), p. 780.
29. *Future Shock*, p. 9.
30. Ibid., p. 429.

Chapter 2—Technology Has Made a Mess of Our Lives

1. "Survey Finds New Yorkers Optimistic on City's Future," *New York Times*, 14 January 1974.
2. "Changes in Social Attitudes . . . ," *New York Times*, 20 August 1973.
3. "The picture of a 'dead' Lake Erie is no longer accurate, according to New York–Pennsylvania Sea Grant Advisory scientists working in the State University of New York system" (David S. Witerski, "Pollution Solutions Revive Lake Erie," Buffalo *Courier-Express*, 2 June 1975). See also William K. Stevens, "Great Lakes Pollution Fight is Gaining," *New York Times*, 23 May 1974, p. 1; and Robert K. Phelps, "The Sonorous Spring," in *New York Times Magazine*, 14 April 1974, p. 14 (re birds returning in increasing numbers).
4. Leonard L. Lederman, director of the National Science Foundation Office of National R & D Assessment, suggests that it may never be possible to do this because we don't have the comparable data for "then" (private communication). Several attempts to get around this problem are reported in E. B. Sheldon and R. Parke, "Social Indicators," *Science*, 16 May 1975, p. 696. The article also contains an extensive bibliography on the subject of social indicators.
5. See, for example, *"The Quality of Life Concept. A Potential Tool for Decision-Makers,"* a report of the Environmental Protection Agency.
6. From *Chemicals and Health*, a study made by the now-defunct President's Science Advisory Committee (Science and Technology Policy Office, National Science Foundation), September 1973, p. 35.
7. See, for example, Alan Anderson, Jr., "The Hidden Plague," *New York Times Magazine*, 27 October 1974.
8. "President's Report on Occupational Safety and Health," 1972. Reported in "The Hidden Plague," by Alan Anderson, Jr., *New York Times Magazine*, 27 October 1974, p. 20.

9. "Traditionally, even the strongest unions have tended to concentrate on winning better wages and to neglect the health and safety needs of workers." From Rachel Scott, "What Work Can Do to You," *New York Times*, 16 September 1974. The article is adapted from her book, *Muscle and Blood, The Massive Hidden Agony of Industrial Slaughter in America* (New York: E. P. Dutton & Co., 1974).

10. See, for example, *Epidemiological Aspects of Carcinogenesis*, pp. 37, 161, 167, and elsewhere. The World Health Organization estimates, nevertheless, that perhaps one-quarter of human cancers result from exposure to environmental agents (not necessarily man-made).

11. *Beard on Bread*, (New York: Knopf, 1973).

12. Beard's book and *Bread*, by Joan Wiener and Diana Collier, Lippincott.

13. Both points are from *The Coming of Post-Industrial Society* (New York: Basic Books, 1973), pp. 161, 162.

14. Ibid., p. 162.

15. Charles A. Reich, *The Greening of America* (New York: Random House, 1970), p. 272.

16. Quoted in Stanley Aronowitz, *False Promises*, (New York: McGraw-Hill, 1973).

17. *New York Times*, 22 March 1974.

18. See William Chapman, "The Blue Collar Blues," *Washington Post*, 14 December 1973; Howard Flieger, "The Other Half," *U.S. News & World Report*, 3 June 1974; "Blue Collar Blues Revisited," *Wall Street Journal*, 7 March 1974; and "There's Still a Car in Your Future," an interview with UAW president Leonard Woodcock, in *Challenge*, May/June 1974.

19. Iradj Siassi, et al, "Loneliness and Dissatisfaction in a Blue Collar Population," *Archives of General Psychiatry*, February, 1974, pp. 261–65.

20. Letter to *New York Times*, 3 July 1974. For more on this subject, see Nancy Foy, "A Tale of Three Factories," *New Scientist*, 10 October 1974.

Chapter 3—The Good Old Days

1. Introduction to Gene Youngblood, *Expanded Cinema,* (New York: Dutton paperback, 1970), p. 32. Anthropologist Morton Klass (Barnard/Columbia) says, however, that it may have seemed that way, but they probably couldn't have managed with less than five to ten thousand words (private communication).

2. A sharp but poignant picture of the lives of the immigrant Jews at this time is painted in *A Bintel Brief,* which is a collection of letters from the Jewish newspaper, the *Daily Forward.* Edited by Isaac Metzger and published by Ballantine, 1972.

3. Don H. Berkebile, "GM, You Can Relax Now," *Smithsonian Magazine,* May 1973.

4. Anthony Wood, *Nineteenth Century Britain, 1815–1914* (London: Longmans, Green & Co., 1960), p. 16.

5. Quoted in *Reports of the United States Commissioners to the Paris Universal Exposition,* 1878, vol. 3, (Washington, D.C.: Government Printing Office, 1880), p. 107.

6. *Arcady, For Better or Worse* (London: T. F. Unwin, 1892), p. 28.

7. Daniel Bell, quoting Jean Fourasié, in *The Coming of Post-Industrial Society* (New York: Basic Books, 1973), p. 189.

8. *Cultural Patterns and Technical Change* (New York: Mentor Books, 1955), p. 5.

9. Quoted in Garrett Hardin, *Stalking the Wild Taboo,* (Los Altos, California: William Kaufman, 1973) p. 196.

10. *Nineteenth Century Britain 1815–1914,* p. 5.

11. Mentioned in H. Muller, *The Children of Frankenstein, A Primer on Modern Technology and Human Values* (Bloomington: Indiana University Press, 1970), p. 9.

12. Quoted in Eugene B. Borowitz's review of *A Passion for Truth,* by A. J. Heschel. In *The New York Times Book Review,* January 13, 1974, p. 22.

13. *Stalking the Wild Taboo,* p. 21.

14. *The Technological Society* (New York: Knopf, 1964), p. 56.
15. *New York Times*, 10 February 1974, Op Ed page.
16. Abraham Maslow refers to the fear of hubris. See *Toward a Psychology of Being* (New York: Van Nostrand Reinhold, 1968), Chapter 5: "The Need to Know and the Fear of Knowing," especially p. 61.
17. See Joseph and Frances Gies, *Life in a Medieval City*, (New York: Apollo Editions, 1973), pp. 114, 115.
18. Quoted in J. D. Bernal, *Science and Industry in the Nineteenth Century* (Bloomington: Indiana University Press, 1970), p. 54.
19. For an eye-opening picture of ancient Rome, see Theodore Crane, "The Squalor That Was Rome," *Natural History*, May 1973.
20. Quoted in Ralph Blumenthal, "Early Files Depict a Grim but Familiar City," *New York Times*, 29 March 1974, p. 37.
21. *Optimism 1* (New York: Norton, 1970), p. 123.
22. Gloria Cole, "Women-Watching in East Africa," Travel Section, *New York Times*, 11 November 1973, p. 1.
23. Review of A. Inkeles and D. H. Smith, *Becoming Modern. Individual Change in Six Developing Countries*, (Cambridge, Mass.: Harvard University Press, 1974). In *Science*, 23 May 1975, p. 829.
24. In the introduction to the 1897–98 edition of Francis Parkman's works. Quoted in Norman Foerster, *American Poetry and Prose* (Cambridge, Mass.: Houghton Mifflin, 1934).
25. *Time* Magazine, September 3, 1973, p. 54.
26. New York: Scribner's, 1973, p. 30.
27. *Human Organization*, Winter 1965, p. 300.
28. Ibid., p. 306.

Chapter 4—Then Why the Technophobe?

1. L. I. Coleman, *Freedom from Fear* (New York: Hawthorn Books, 1954), p. 28.
2. M. L. Aronson, *How to Overcome Your Fear of Flying,*

(New York: Warner Paperback Library, 1973), p. 61.

3. *The Ordeal of Change* (New York: Harper & Row, 1952; rev. ed., 1963), p. 1.

4. "A Dream of Waters Glittering With Stars," *Impact of Science on Society*, October–December 1969, p. 357.

5. John Lukacs, "It's Halfway to 1984," *New York Times Magazine*, 2 January 1966.

6. *Freedom from Fear*, p. 38.

7. "The Genesis of Pollution," *New York Times*, 16 September 1973, (Op Ed page); excerpt from *Horizon* Magazine.

8. (Urbana: University of Illinois Press, 1970), p. 16.

9. *The Coming of Post-Industrial Society* (New York: Basic Books, 1973), p. 29.

10. *Bulletin of the Atomic Scientists*, June 1973, p. 9.

11. *Bulletin of the Atomic Scientists*, May 1971, p. 4.

12. Erwin Chargaff, "The Paradox of Biochemistry," *Columbia University Forum*, Summer 1969.

13. Leonard L. Lederman (see note 4 to Chapter 2), using estimates of Griliches, Mansfield, Minasian, and others, suggests that the figure is about 40 or 50 percent. These estimates and a further discussion of the question can be found in *Research and Development and Economic Growth/Productivity*, NSF 72–303, 1972. See also Edwin Mansfield, *The Economics of Technological Change*, (New York: Norton, 1968).

14. *Expanded Cinema*, New York: Dutton paperback, 1970, p. 418.

Chapter 5—Social Progress and Technology

1. "A Dream of Waters Glittering With Stars," *Impact*, October/December 1969, p. 359.

2. Robert Gillette, "Science in Mexico (I)," *Science*, 15 June 1973, p. 1154.

3. Some writers dispute this, however. See, for example, Richard Parker, *The Myth of the Middle Class* (New York: Liveright, 1972).

4. L. Rocks and R. P. Runyon, *The Energy Crisis*, (New York: Crown Publishers, 1972), p. 12.
5. *Traditional Cultures: And the Impact of Technological Change* (New York: Harper & Row, 1962), p. 48.
6. See P. M. Worsley, "Cargo Cults," *Scientific American*, May 1959.
7. Ibid., pp. 122–26.
8. An example of aid being given for the wrong reasons is the Yir Yiront tribe of Australia. The whole sad story is given in Lauriston Sharpe's classic paper, "Steel Axes for Stone-Age Australians," *Human Organization*, 2 (1952): 17–22. Reprinted as "Technological Innovation and Culture Change: An Australian Case," in Peter B. Hammond, ed., *Cultural and Social Anthropology, Selected Readings*, (New York: Macmillan, 1964), pp. 84–94.
9. In "The Evolutionary Crisis," *The Futurist*, February 1975, p. 10.
10. *Traditional Cultures*, p. 171.
11. In *Cultural and Social Anthropology*, p. 93.

Chapter 6—The Turn to Irrationalism

1. William J. Peterson, *Those Curious New Cults* (New Canaan, Conn.: Keats Publishing, 1973), p. 14.
2. "Gallup Poll Indicates 32 Million Believe in Astrology," *New York Times*, October 19, 1975, p. 46. The article also reports that 1,250 out of the nation's 1,500 daily newspapers carry astrology columns.
3. Peterson, *Those Curious New Cults*, p. 14.
4. Ibid., p. 47.
5. Ibid., p. 62.
6. Ibid., p. 87.
7. Boyce Rensberger, "Evolution Theory Still Disputed 50 Years After 'Monkey Trial,'" 10 July 1975, p. 58.
8. Ibid.
9. William T. Daly, associate professor of political science at Stockton State College in New Jersey, maintains that

Hitler and nazism were an antitechnological response to the problems faced by Germany after World War I. The idea is developed in his forthcoming book, *America and Her Critics* (New York: Alfred Publishers, Inc.)

10. *Pensées.* See XIV, no. 894.
11. "The Nature and Sources of Irrationalism," *Science*, 1 June 1973, p. 928. For a current example of this myth, see Theodore Roszak, "The Monster and the Titan: Science, Knowledge and Gnosis," *Daedalus*, Summer 1974, pp. 17–32.
12. See, for example, Richard Hofstadter, *Anti-Intellectualism in American Life*, (New York: Alfred A. Knopf, 1963).
13. *The Greening of America* (New York: Random House, 1970), p. 261.
14. Ibid., p. 256.
15. Quoted in Thomas Meehan, "The Flight From Reason," *Horizon*, Spring 1970, p. 10.
16. *Where the Wasteland Ends* (Garden City, N.Y.: Doubleday, 1972), pp. 78–87.
17. Quoted in Steve Aaronson, "Pictures of an Unknown Aura," *The Sciences*, January/February 1974, p. 17.
18. Ibid., p. 15 and cover.
19. Letter to *Science*, 8 November 1974, p. 480.
20. "The Role of the Spirit in creating the Future Environment," in *Environment and Change: The Next Fifty Years*, (Bloomington: Indiana University Press, 1968), p. 58.
21. See, for example, Roberto Vacca, *The Coming Dark Age* (Garden City, N.Y.: Doubleday, 1973); Robert Heilbroner, *An Inquiry Into the Human Prospect*, (New York: Norton, 1974; J. A. Livingston, *One Cosmic Instant. Man's Fleeting Supremacy* (Boston: Houghton Mifflin, 1973); and Philip Slater, *Earthwalk*, (Garden City, N.Y.: Doubleday, 1974).
22. Quoted in Z. Brzezinski, *Between Two Ages: America's Role in the Technetronic Era* (New York: Viking, 1970), p. 96.

Chapter 7—Technofears—and What About Them?

1. New York: Harper & Row, 1972, p. 48.
2. Ibid., p. 135.
3. Quoted in Seymour Melman, "After the Military-Industrial Complex?" *Bulletin of the Atomic Scientists*, March 1971, p. 7.
4. Ibid., p. 8. Melman is professor of industrial and management engineering at Columbia University.
5. William Peterson, *Those Curious New Cults* (New Canaan, Conn.: Keats Publishing Co., 1973), p. 5.
6. See, for example, George Gamow, *Thirty Years That Shook Physics, The Story of Quantum Theory* (Garden City, N.Y.: Doubleday, 1966); Barbara Lovett Clive, *The Questioners*, Physicists and the Quantum Theory (New York: Crowell, 1965); or Werner Heisenberg, *Physics and Beyond: Encounters and Conversations* (New York: Harper & Row, 1970).
7. See, for example, A. D. Gordon, *Selected Essays* (New York: Arno, 1938). Gordon was one of the early Jewish pioneers who helped in establishing Israel.
8. *New York Times Magazine*, 11 August 1974, p. 10.
9. Joan A. Rothschild, "A Feminist Perspective on Technology and the Future of Human Society." (Paper presented at the Second General Assembly of the World Future Society, June 2–5, 1975), p. 7. See also Shulamith Firestone, *The Dialectic of Sex: The Case for a Feminist Revolution* (New York: Morrow, 1970), and Ursula Le-Guin, *The Left Hand of Darkness* (New York: Ace Books, 1969).
10. September 1973, pp. 584 and 586.
11. *Lancet*, August 1972, p. 422.
12. The story is beautifully told by the brain surgeon himself, Dr. I. S. Cooper, in his book, *The Victim is Always the Same*, (New York: Harper & Row, 1973.
13. "The Slippery Slope of Science," *Science*, 15 March 1974,

p. 1041. For additional discussion, see his book, *Genetic Fix* (New York: Macmillan, 1973).

14. See Gina Bari Kolata, "Freedom of Information Act: Problems at the FDA," *Science*, 4 July 1975, pp. 32, 33.
15. *New Scientist*, 30 May 1974, p. 554.
16. *The Domination of Nature* (New York: Braziller, 1972), p. 11.
17. See David C. Anderson, "Can Supertechnology Bring Peace?" *Wall Street Journal*, 22 June 1970, and Bruce Callander, "War Control," *Air Force Times*, 28 March 1973.
18. "Letters," *Science*, 1 December 1972, p. 933.
19. *Public Interest Report*, January 1975, p. 5. For more on this question see John McPhee, *The Curve of Binding Energy* (New York: Farrar, Straus & Giroux, 1974), and Mason Willrich and Theodore B. Taylor, *Nuclear Theft: Risks and Safeguards* (Cambridge, Mass.: Ballinger, 1974).
20. It is reported in *Science News*, 7 December 1974, that gas-core reactors might be used to destroy radioactive wastes. A high neutron flux can transmute them into isotopes that either decay faster or not at all (p. 362).
21. Speech at Monadnock Summer Lyceum, Monadnock, N.H., 30 July 1972.

Chapter 8—Time To Cut Down?

1. *The Next Hundred Years. The Unfinished Business of Science* (New York: Reynal & Hitchcock, 1936), p. 215.
2. *Energy Crises in Perspective* (New York: Wiley, 1974), p. 20. Since 1850 our per capita use has only doubled. "Many people," Fisher adds, "think that we used much less energy in 1850 because they forget wood (which heated most of our homes and commercial buildings) and they forget hay and oats (which fueled our transportation system) (private communication, Jan. 3, 1975).
3. " 'International' Technology and the U.S. Economy: Is

There a Problem?" in *The Effects of International Technology Transfers on U.S. Economy*, National Science Foundation, July 1974, p. 62.

4. "Can Technology Be Humanized—in Time?" *National Parks*, July 1969, p. 4. For more on this, see Steven Weinberg, "Reflections of a Working Scientist," in *Daedalus*, Summer 1974, p. 33. Alvin Weinberg, director of Oak Ridge National Laboratory, proposed a set of criteria, including social criteria, to help decide how to split government support among the various scientific disciplines. The criteria can also be used to help evaluate science as an activity of society. A. Weinberg, *Reflections on Big Science* (Cambridge, Mass.: MIT Press, 1967), chap. 4.

5. For a picture of a society that tried to do this, see Samuel Butler's *Erewhon*.

6. *First Things, Last Things* (New York: Harper & Row, 1971), p. 29.

7. In "Rich Countries and Poor in a Finite, Interdependent World," *Daedalus*, Fall 1973, p. 162.

8. Rudolph Klein, "Growth and Its Enemies," *Commentary*, June 1972, p. 43.

9. The big military nations spend $230 billion annually on arms. ("Logical—Not Zero—Growth," C. L. Sulzberger, *New York Times* 11 January 1975. Op Ed page).

10. See, for example, D. L. Meadows, et al, *The Limits to Growth* (New York: Universe Books, 1972), pp. 110, 111.

11. Wilfred Malenbaum, "World Resources for the Year 2000," *Annals of the American Academy of Political and Social Scence*, July 1973, p. 35.

12. Figure supplied by Leonard L. Lederman (private communication).

13. Quoted in H. J. Maidenberg, "Resources: A Calm Voice," *New York Times*, 12 May 1974, p. 5 (Business and Finance section).

14. In "On Reforming Economic Growth," *Daedalus*, Fall 1973, p. 125.

15. See Philip Kotler and S. J. Levy, "Demarketing, Yes, De-marketing," *Harvard Business Review*, November/December 1971; also by the same authors, "Demarketing: How to Reduce the Demand," *New York Times*, June 9, 1974, Op Ed page.

16. See Hazel Henderson, "The Decline of Jonesism," *The Futurist*, October 1974.

17. "Some observers have noted that productivity has not risen much more since World War II than before the war, despite significant increases in R & D, and therefore question the contribution of R & D to increased pr ductivity. It is important to remember that a large part of R & D investment in the United States has been for defense and space objectives. Such R & D investments are essentially for purposes other than economic ones, and their contributions to economic growth relative to their cost is limited, despite some significant spillovers to civilian technology (e.g., computers, atomic energy)" (Leonard L. Lederman, "The Socio-Economic Implications of Science and Technology," [Paper presented at the Exxon Research and Engineering Company Seminar, 23 October 1974], p. 11).

18. Council of Economic Advisors, *Annual Report*, 1964, p. 92.

19. From U.S. Department of Commerce figures. See Michael Boretsky, "Trends in U.S. Technology: A Political Economist's View," *American Scientist*, January/February 1975, p. 80. R. B. Stobaugh of Harvard University points out that if management fees are included, the figure leaps to thirteen times as high ("Summary and Assessment of Research Findings of U.S. International Transactions Involving Technology Transfers," in *The Effects of International Technology Transfers on U.S. Economy*, National Science Foundation (NSF 74–21), pp. 15 and 93.)

20. For more on Boretsky's ideas, see his *U.S. Technology: Trends and Policy Issues*, Monograph no. 17, George Washington University, October 1973 (available from National

Technical Information Service), and the *American Scientist* article mentioned in note 19 above. His views are also given in Deborah Shapley, "Technology and the Trade Crisis: Salvation Through a New Policy?" *Science*, 2 March 1973. See also Sherman Gee, "Foreign Technology and the United States Economy," *Science*, 21 February 1975.

21. "Trends in U.S. Technology," pp. 70 and 71.
22. Quoted in *The Futurist*, February 1972, p. 17.

Chapter 9—Technology and the Citizen

1. *Scientific American*, May 13, 1899, p. 293.
2. Eric Hoffer, *Last Things, First Things* (New York: Harper & Row, 1971), p. 99.
3. See, for example, A. L. Hammond and T. H. Maugh III, "Stratospheric Pollution: Multiple Threats to Earth's Ozone," *Science*, 25 October 1974, p. 335; Allen L. Hammond, "Ozone Destruction: Problem's Scope Grows, Its Urgency Recedes," *ibid.*, 28 March 1975, p. 1183; and N. E. Hester, et al., "Fluorocarbon Air Pollutants: Measurements in the Lower Stratosphere," *Environmental Science & Technology*, September 1975, p. 875.
4. See B. J. Culliton, "The Politics of Biology: Young Academicians Becoming Involved," *Science*, 26 April 1974, p. 445.
5. See, for instance, Henry Lansford, "Weather Modification: the Public Will Decide," *Bulletin of the American Meteorological Society*, July 1973, pp. 568–70; and Luther J. Carter, "Weather Modification: Colorado Heeds Voters in Valley Dispute," *Science*, 29 June 1973, pp. 1347–50. Also see B. J. Culliton, "Kennedy: Pushing for More Public Input in Research," *Science*, 20 June 1975, pp. 1187–89.
6. In "The Waning Promise of Nuclear Power," *Science Digest*, February 1975, p. 75.
7. "Proposed New Energy Technologies off the Shore of New

Jersey and Delaware" (booklet), Office of Technology Assessment, p. 9.

8. For more on this, see J. G. Mitchell and C. L. Stallings, eds., *Ecotactics: The Sierra Club Handbook for Environmental Activists* (New York: Pocket Books, 1970).

9. Ibid., p. 32.

10. *New York Times*, 3 August 1975, editorial page. In all fairness it should be pointed out that the Army Corps of Engineers was acting at the behest of Congress, and only after severe flooding in 1955 had caused ninety-nine deaths and millions of dollars in property damage. See Donald Janson, "Tocks Dam: Story of 13-Year Failure," *New York Times*. 4 August 1975, p. 23.

11. Roy Reed, "Postmark Names Returning After an Editor's Crusade," *New York Times*, 19 August 1975.

12. As reported in the book *Whistle Blowing*, edited by Ralph Nader, et al. (New York: Bantam Books, 1972).

13. See Nicholas Wade, "Protection Sought for Whistle Blowers," *Science*, 7 December 1973, p. 1002.

14. B. J. Culliton, "National Research Act: Restores Training, Bans Fetal Research," *Science*, 2 August 1974, pp. 426, 427; Peter Frishouf, "End of the Free Ride," *New Physician*, August 1974, pp. 14–19; and Wendy Bone, "The Price of a Free Ride," *New Physician*, April 1975, pp. 36–38.

15. Others agree. See, for example, Benjamin S. P. Shan, "Science Literacy: the Public Need," *The Sciences*, January/February 1975, pp. 27, 28.

16. *New Scientist*, 5 July 1973, p. 15.

17. Personal interview, October 29, 1973.

18. As Z. Brzezinski puts it: "How to combine social planning with personal freedom is already emerging as the key dilemma of technetronic America, replacing the industrial age's preoccupation with balancing social needs against requirements of free enterprise." (*Between Two Ages: America's Role in the Technetronic Era* (New York: Viking, 1970), p. 260.

Chapter 10—Sociotechnological Needs

1. In *The Art of Conjecture* (New York: Basic Books, 1967), p. 282.
2. See Wilson Clark, "It Takes Energy to Get Energy; the Law of Diminishing Returns Is in Effect," *Smithsonian Magazine*, December 1974, pp. 84–90.
3. A 1970 study by the U.S. Bureau of Mines showed the following injury rates per million man hours of work:

All manufacturing	15.2
Coal surface mining	30.4
Heavy construction	32.7
Coal underground mining	53.0
Oil and gas well drilling	69.8

   Source: F. T. Moyer and M. B. McNair, "Injury Experience in Coal Mining, 1970," U.S Bureau of Mines Information Circular 8613, 1973.
4. Dr. George O. G. Löf, president of the International Solar Energy Society and head of the Solar Energy Department at Colorado State University, estimates the cost at between $5,000 and $10,000 for a typical solar home heating system, depending on size and location. He reports: "The Solaron Corp., the only company marketing a complete system for adequate solar heating of buildings, has numerous contracts below $5,000, and, on the other hand, a few above $10,000." (Private communication, September 5, 1975.)
5. See, for example, T. B. Reed and R. M. Lerner, "Methanol: A Versatile Fuel for Immediate Use," *Science*, 28 December 1973.
6. In "Firewood," *New Yorker*, 25 March 1974, p. 105.
7. While fuel consumption will be higher for methanol than for gasoline on a weight or volume basis, T. B. Reed and R. M. Lerner maintain that "specific energy consumption (energy per kilometer) will certainly be lower because higher compression ratios and simpler pollution controls

can be used" (Methanol: A Versatile Fuel,") p. 1301.

8. In "Methanol as a Gasoline Extender: A Critique," *Science*, 29 November 1974, p. 785.

9. Dr. George O. G. Löf estimates that one hundred to two hundred homes in the United States are now partially heated by solar energy. ("Rapid Gains Seen for Solar Energy." Gladwin Hill, *New York Times*, 3 August 1975, p. 43.)

10. See Michael Harwood, "Energy from our Star Will Compete with Oil, Natural Gas, Coal and Uranium. But Not Soon," *New York Times Magazine*, 16 March 1975, p. 42.

11. See W. C. Gough and B. J. Eastlund, "Energy, Wastes and the Fusion Torch," U.S. Atomic Energy Commission, 27 April 1971.

12. For more on these processes, see H. Hellman, *Energy in the World of the Future* (New York: Evans, 1973).

13. March 1967, p. 121.

14. See, for example, E. F. Lindsley, "First Report on Smokey Yunick's Total Energy System," by *Popular Science*, August 1975. ("Wind and sun will heat, cool, and make electricity, liquid fuel and fertilizer.")

15. *Exploring Energy Choices*, Energy Policy Project of the Ford Foundation, 1974.

16. See M. T. Farvar and J. P. Milton, *Careless Technology— Ecology and International Development* (New York: Natural History Press, 1972).

17. John McDermott, "Intellectuals and Technology," *New York Review of Books*, July 31, 1969, p. 29.

18. *Whole Earth Catalog* (Menlo Park, California: Portola Institute, 1970), p. 23.

19. Personal interview, August 30, 1973.

20. For more on the intermediate technology approach, see Nicholas Wade, "E. F. Schumacher: Cutting Technology Down to Size," *Science*, 18 July 1975, p. 199. See also Schumacher's book, *Small Is Beautiful* (New York: Harper & Row, 1973), and articles by and about him in *The Futurist*, December 1974, pp. 274–84. Another interesting

article in the same issue is J. P. Milton, "Communities That Seek Peace With Nature," pp. 264–69. See also Nicholas Wade, "New Alchemy Institute: Search for an Alternate Agriculture," *Science*, 28 February 1975, pp. 727–29.

21. The Clivus Multrum is being marketed in the United States by Clivus Multrum USA, 14A Eliot Street, Cambridge, Mass. 02138. Or see "Clivus Multrum: the Careful Technology," *The Futurist*, December 1974, pp. 270–71.

22. W. E. Westman and R. M. Gifford, "Environmental Impact: Controlling the Overall Level," *Science*, 31 August 1973, pp. 819–24.

23. "Letters," *Science*, 28 December 1973, p. 1296.

24. Robert Bendiner reports that during the depression of the 1930s, a group called the technocrats "were for letting the engineers run the country, with the price of all commodities and services recalculated in . . . units of energy . . ." ("The Depression. How It Really Was," *New York Times Magazine*, pp. 92 ff).

25. "As recently as the late thirties," says Lester Brown, president of Worldwatch Institute and coauthor of *By Bread Alone*, "each of the continents except Western Europe was a net exporter of grain. . . . A generation later, the world pattern of grain trade has changed dramatically. . . . All major regions are now importing grain except North America, which remains with Australia as the only net exporter." ("Food: The Next Quarter Century," [Paper presented at Second General Assembly of the World Future Society, June 2, 1975]).

26. See, for example, John H. Douglas, "Climate Change: Chilling Possibilities," *Science News*, 1 March 1975, pp. 138–140, and Tom Alexander, "Ominous Changes in the World's Weather," *Fortune*, February 1974, pp. 90–95. For an opposing view, see Wallace S. Broecker, "Climate Change: Are We on the Brink of a Pronounced Global Warming?" *Science*, 8 August 1975, pp. 460–63.

27. Gladwin Hill, "Population Boom and Food Shortage," *New York Times*, 17 August 1975, p. 35. For more on the Green Revolution, see Nicholas Wade, "Green Revolution: A Just Technology, Often Unjust in Use," *Science*, pt. I, 20 December 1974, pp. 1093–96; pt. II, 27 December 1974, pp. 1186–92.

28. How much of a factor is a disputed point, however. C. W. McMillan, executive vice president of the American National Cattlemen's Association, says, "Grain represented only 19 percent of the feed eaten by beef cattle in 1973–74. And those grains were not food grains but coarse feed grains. . . . At most only about 50 percent of our feed grain is of a grade or type suitable for processing into human food." Lester Brown, president of Worldwatch Institute, counters: "The fact is that all cereals are edible; indeed each of the seven cereals now grown in the world—wheat, rice, corn, sorghum, barley, rye and oats—were all domesticated as foodgrains. It has only been quite recently, in historical terms, that substantial quantities of any cereals have been fed to cattle. Each of the cereals is an important staple or substaple some place in the world. Corn, which is primarily used as a feedgrain in the United States, is the dominant food staple in several countries in Africa and Latin America" ("Meat Mythology—the Role of Animals in Future World Food Supply," C. W. McMillan, *Vital Speeches of the Day*, June 15, 1975, p. 532; Lester Brown, "Food: the Next Quarter Century" [Paper presented at the Second General Assembly of the World Future Society, June 2, 1975], p. 5.

29. In the magazine *Agricultural Research*, published by the U.S. Department of Agriculture, the following point is made: ". . . such ruminants as cattle and sheep are superbly endowed to thrive on forages—pasture and harvested herbage—converting fibrous material that people cannot eat into protein-rich meat and milk. Indeed, forages account for about 70 percent of the nutrients that beef

cattle consume over their lifetimes. This is a notable statistic because over half the total U.S. land area—about a billion acres—is fit not for cropping but for producing forage" (May 1974, p. 2).

30. D. D. Rutstein, statement to the Committee on Ways and Means, House of Representatives, June 21, 1974.

31. Jean L. Marx, "Nitrogen Fixation in Maize," *Science*, 1 August 1975, p. 368.

32. "Living with Pests," by Mary-Jane Schneider, *The Sciences*, December 1973.

33. See the section on toxic food in Dr. Manocha's book *Nutrition and Our Overpopulated Planet*, (Springfield, Illinois: Charles C. Thomas, 1975).

34. The Food and Drug Administration reports a death rate of 150 per million for "no method of contraception"! That is, ... out of 100,000 women not using contraceptives between the ages of 15 and 44, 60,000 will become pregnant. Of these 60,000 there will be 15 deaths" (Mary-Carol Kelly, FDA: Private communication, August 15, 1975).

35. "Heat in Male Contraception (Hot Water 60°c, Infrared, Ultrasound, and Microwaves)" (Paper presented at the 58th Annual Meeting of the Federation of American Societies for Experimental Biology, April 10, 1974), p. 1. See also Clive Wood, "The Hazards of Vasectomy," *New Scientist*, 3 May 1973.

36. Fahim, Hall, and Der, "Heat in Male Contraception," pp. 3, 4.

37. "Wool Triumphs Over Flames," *Agricultural Research* (U.S. Department of Agriculture), December 1974, pp. 3, 4.

38. Teleoperators are devices which can mimic human activities, generally under human control. It is possible to include force feedback so that the operator can "feel" what is going on. See William R. Corliss, "Teleoperators: Man's Machine Partners," U.S. Energy Research and Development Administration, 1972.

39. "Medical-Care Gap Cut by Students," *New York Times,* August 12, 1974, p. 49.

40. See, for example, Hal Hellman, *Biology in the World of the Future* (New York: Evans, 1971), especially chap. 2.

41. In "Evolving New Methods for Health Care Delivery" (Paper presented at the 141st Annual Meeting of the American Association for the Advancement of Science, New York, January 27, 1975), p. 3.

42. *People of the United States in the 20th Century,* I. B. Taeuber & Co., U.S. Department of Commerce, December 1971, pp. 328, 329.

43. "Can Swamps and Civilization Co-Exist?" Anthony Wolff, *RIF Illustrated* (Rockefeller Foundation), November 1973, p. 5.

44. There is some small evidence of a change in this trend. Author William Ellis reports that between 1970 and 1973 eighteen of the twenty fastest growing states were rural ones! (In a paper presented at the Second General Assembly of the World Future Society, Washington, D.C., June 4, 1975.)

45. Peter Goldmark and Bonnie Kraig, "Technology in the Small Town." Paper presented at the 141st Annual Meeting of the American Association for the Advancement of Science, January 28, 1975.

46. Michael Knight, "Scientist Is Testing New Rural Society Based on Electronics," *The New York Times,* October 28, 1973, p. 57.

47. In "What's Human About Man is His Technology," *Smithsonian Magazine,* May 1973, p. 28.

48. John Clare, "How EMR [Department of Energy, Mines, and Resources] Works to Save You Fuel," *GEOS* (Canada), Fall 1974, p. 12.

49. Personal interview, 27 September 1973.

50. In *So Human an Animal* (New York: Scribner's, 1968), p. xi.

Chapter 11—Fear and the Future

1. *Modern Man Is Obsolete* (New York: Viking Press), 1945. (I am not convinced that this would prolong his existence on this planet, however.)
2. In *New Reformation: Notes of a Neolithic Conservative* (New York: Random House, 1970), p. 20. Thorstein Veblen once characterized American society as a residue of barbarism in the setting of modern technology.

BIBLIOGRAPHY

Books (Nonfiction)

ADLER, CY A. *Ecological Fantasies, Death From Falling Watermelons.* New York: Green Eagle Press, 1974.

ALEXANDER, FRANZ. *Our Age of Unreason.* Philadelphia: J. B. Lippincott Company, 1942.

AMERICAN ACADEMY OF ARTS AND SCIENCES. "The No-Growth Society," *Daedalus,* Fall 1973.

ARMYTAGE, W. H. G. *The Rise of the Technocrats.* London: Routledge and Kegan Paul, 1965.

ARON, RAYMOND, ed. *World Technology and Human Destiny.* Ann Arbor: University of Michigan Press, 1963.

ARONSON, MARVIN L. *How to Overcome Your Fear of Flying.* New York: Hawthorn Books, 1973. Warner Paperback Library edition, 1973.

BACON, FRANCIS. *New Atlantis.* Various editions.

BAGDIKIAN, BEN. *The Information Machines: Their Impact on Men and the Media.* New York: Harper & Row, Publishers, 1971.

BAIER, KURT, and RESCHER, NICHOLAS, eds. *Values and the Future: The Impact of Technological Change on American Values.* New York: Free Press, 1969.

BARBER, BERNARD. *Science and the Social Order.* New York: Collier Books, 1962.

BARNETT, H. G. *Innovation: The Basis of Cultural Change.* New York: McGraw-Hill, 1953.

BARNETT, H., and MORSE, C. *Scarcity and Growth.* Baltimore: Johns Hopkins University Press, 1963.

BARRETT, WILLIAM. *Irrational Man.* London: Mercury Books, 1964.

BARZUN, JACQUES. *Science, the Glorious Entertainment.* New York: Harper & Row, 1964.

BECKER, ERNEST. *The Denial of Death.* New York: Free Press, 1973.

BELL, DANIEL. *The Coming of Post- Industrial Society, A Venture in Social Forecasting.* New York: Basic Books, 1973.

————, ed., *Toward the Year 2000: Work in Progress.* Boston: Houghton Mifflin Company, 1968.

BENJAMIN, A. CORNELIUS. *Science, Technology and Human Values.* Columbia: University of Missouri Press, 1965.

BERNAL, J. D. *Science and Industry in the Nineteenth Century.* Bloomington, Ind.: Indiana University Press, 1953, rev. ed., 1970.

————. *The Social Function of Science.* Cambridge, Mass.: MIT Press, 1967.

————. *The World, the Flesh and the Devil.* Bloomington: Indiana University Press, 1969.

BERRY, ADRIAN. *The Next Ten Thousand Years. A Vision of Man's Future in the Universe.* New York: Saturday Review Press, 1975.

BETTMAN, O. L. *The Good Old Days—They Were Terrible.* New York: Random House, 1974.

BOGUSLAW, ROBERT. *The New Utopians.* Englewood Cliffs, N.J.: Prentice-Hall, 1965.

BORGESE, ELISABETH MANN. *The Ascent of Woman.* New York: George Braziller, 1962.

BOYKO, HUGH, ed. *Science and the Future of Mankind.* Bloomington: Indiana University Press, 1961.

BRADEN, WILLIAM. *The Age of Aquarius: Technology and the*

Cultural Revolution. New York: Quadrangle Books, 1970.

BRONOWSKI, JACOB. *Science and Human Values*. New York: Julian Messner, 1956.

BRONWELL, A. B. *Science and Technology in the World of the Future*. New York: Wiley-Interscience, 1970, 1971.

BROSSEAU, R., ed. *Looking Forward*. New York: American Heritage Publishing Company, 1970.

BRZEZINSKI, ZBIGNIEW. *Between Two Ages: America's Role in the Technotronic Era*. New York: Viking Press, 1970.

BURKE, J. G. *The New Technology and Human Values*. 2nd ed. Belmont, California: Wadsworth, (paperback), 1972.

BURNS, J. M. *Uncommon Sense*. New York: Harper & Row, Publishers, 1971.

BUSH, VANNEVAR. *Science Is Not Enough*. New York: William Morrow and Company, 1967.

CALDER, NIGEL. *Eden Was No Garden, an Inquiry into the Environment of Man*. New York: Holt, Rinehart and Winston, 1967.

———. *Technopolis. Social Control of the Uses of Science*. New York: Simon and Schuster, 1969, 1970.

CARDWELL, D. S. L. *Technology, Science and History*. London: William Heinemann, 1972.

———. *Turning Points in Western History*. New York: Science History Publications paperback, 1972.

CARLETON, R. M. *False Prophets of Pollution; How Fake Ecologists Sidetrack America's Progress*. Tampa, Florida: Trend House, 1973.

CARPENTER, EDWARD. *Forecasts of the Coming Century*. Manchester: Labour Press, 1897.

CARROLL, C. M. *The Great Chess Automaton*. New York: Dover, 1975.

CATTELL, R. B. *A New Morality From Science: Beyondism*. New York: Pergamon Press, 1973.

CETRON, M. J., and BARTOCHA, B., eds. *The Methodology of Technology Assessment*. New York: Gordon and Breach paperback, 1972.

CHAPLIN, G., and PAIGE, G. D., eds. *Hawaii 2000*. Honolulu: University Press of Hawaii, 1973.

CLARKE, A. C. *Profiles of the Future*. Rev. ed. New York: Harper & Row, Publishers, 1973.

COLEMAN, L. I. *Freedom from Fear*. New York: Hawthorn Books, 1954.

COLUMBIA UNIVERSITY, SEMINAR ON TECHNOLOGY AND SOCIAL CHANGE. *The Impact of Science on Technology*. New York: Columbia University Press, 1965.

COMMONER, BARRY. *Science and Survival*. New York: Viking Press, 1967.

COOPER, I. S. *The Victim Is Always The Same*. New York: Harper & Row, Publishers, 1973.

COOPER, PAULETTE. *The Scandal of Scientology*. New York: Tower Publications, 1972.

COTE, A. J., JR. *The Search for the Robots*. New York: Basic Books, 1967.

COWLES, E. S. *Don't Be Afraid! How to Get Rid of Fear and Fatigue*. New York: McGraw-Hill, 1941.

COYLE, D. L. *Waste*. Indianapolis: Bobbs-Merrill Company, 1936.

DALKAY, N. C., with ROURKE, D. L., et. al. *Studies in the Quality of Life: Delphi and Decision-Making*. Boston: D. C. Heath & Company, 1974.

DELGADO, J. M. R. *Physical Control of the Mind*. New York: Harper & Row, Publishers, 1969.

DERRY, T. K., and WILLIAMS. T. I. *A Short History of Technology*. New York: Oxford University Press, 1961.

DIAMOND, SIGMUND, ed. "Science in Human Affairs," special issue of the *Proceedings of The Academy of Political Science*, April 1966.

DIEBOLD, JOHN. *Man and the Computer; Technology as an Agent of Social Change*. New York: Praeger Publishers, 1969.

DOBZKANSKY, THEODOSIUS. *The Biology of Ultimate Concern*. New York: New American Library, 1967.

DORF, R. C. *Technology, Society and Man*. San Francisco: Boyd and Fraser Publishing Co., 1974.

DUBOS, RENE. *A God Within.* New York: Charles Scribner's Sons, 1972.

———. *Reason Awake: Science for Man.* New York: Columbia University Press, 1970.

———. *So Human an Animal.* New York: Scribner's, 1968.

ECO, V., and ZORZOLI, G. B. *The Picture History of Inventions.* New York: Macmillan Company, 1963.

EISELEY, LOREN. *The Man Who Saw Through Time.* New York: Charles Scribner's Sons, paperback, 1973.

ELDRIDGE, H. W. *The Second American Revolution.* New York: William Morrow and Company, 1964.

ELLUL, JACQUES. *A Critique of the New Commonplace.* New York: Alfred A. Knopf, 1968.

———. *The Technological Society.* New York: Alfred A. Knopf, 1964.

EMERSON, EDWIN, JR. *A History of the Nineteenth Century, Year by Year.* New York: P. F. Collier and Son, 1902.

ESFANDIARY, F. M. *Optimism I.* New York: W. W. Norton & Company, 1970.

———. *Upwingers.* New York: John Day Company, 1973.

EURICH, NELL. *Science in Utopia.* Cambridge, Mass.: Harvard University Press, 1967.

EWALD, W. R., JR., ed. *Environment for Man.* Bloomington: Indiana University Press, 1967.

FARVAR, M. T., and MILTON, J. P. *Careless Technology—Ecology and International Development.* New York: Natural History Press, 1972.

FEINBERG, GERALD. *The Prometheus Project.* Garden City, N.Y.: Doubleday & Company, 1969.

FERKISS, V. *The Future of Technological Civilization.* New York: George Braziller, 1974.

———. *Technological Man: The Myth and the Reality.* New York: George Braziller, 1969.

FIEDLER, L. A. *Waiting for the End.* New York: Delta Books, 1964.

FISHER, J. C. Energy Crisis in Perspective. New York: John Wiley and Sons, 1974.

FORBES, R. J. The Conquest of Nature: Technology and Its Consequences. New York: Praeger Publishers, 1968.

FOSTER, G. MCC. Traditional Cultures: and the Impact of Technological Change. New York: Harper, 1962.

FRANKLIN, H. B. Future Perfect. New York: Oxford University Press, 1966.

FREEDLAND, NAT. The Occult Explosion. New York: G. P. Putnam's Sons, 1972.

FREEDMAN, JONATHAN. The Psychology of High Density Living. New York: The Viking Press, 1975.

FULLER, R. B. Operating Manual for Spaceship Earth. Carbondale: Southern Illinois University Press, 1969.

———. Utopia or Oblivion: The Prospects for Humanity. New York: Bantam Books, 1969.

FURNAS, C. C. The Next Hundred Years. The Unfinished Business of Science. New York: Reynal & Hitchcock, 1936.

GALBRAITH, J. K. The Affluent Society. 2nd ed. Boston: Houghton Mifflin Company, 1969.

———. Economics and the Public Purpose. Boston: Houghton Mifflin Company, 1973.

———.The New Industrial State. Boston: Houghton Mifflin Company, 1967.

GARDNER, MARTIN. Fads and Fallacies in the Name of Science. New York: Dover Publications, 1957.

GAY, PETER. The Enlightenment: An Interpretation. 2 vol. New York: Alfred A. Knopf, 1966–69.

GERLACH, L. P., and HINE, V. H. Lifeway Leap: The Dynamics of Change in America. Minneapolis: University of Minnesota Press, 1973.

GIDEON, SIGFRIED. Mechanization Takes Command. New York: Oxford University Press, 1948.

GIES, J., and GIES, F. Life in a Medieval City, New York: Apollo Editions, 1969; paperback, 1973.

GILKEY, LANGDON. Religion and the Scientific Future, Reflections *on Myth, Science, and Theology.* New York: Harper & Row, 1970.

GILMAN, WILLIAM. *Science: U.S.A.* New York: Viking Press, 1965.

GINSBERG, ELI, ed. *Technology and Social Change.* New York: Columbia University Press, 1964.

GLASS, BENTLEY. *Science and Ethical Values.* Chapel Hill: University of North Carolina Press, 1965.

GOODMAN, PAUL. *New Reformation: Notes of a Neolithic Conservative.* New York: Random House, 1970.

GOODY, JACK. *Technology, Tradition, and the State in Africa.* New York: Oxford University Press, 1971.

GORDON, A. D. *Selected Essays.* New York: Arno, 1938.

GOULDNER, A. W., and PETERSON, R. A. *Notes on Technology and the Moral Order.* Indianapolis: Bobbs-Merrill Company, 1962.

GRAVES, ROBERT. *Difficult Questions, Easy Answers.* Garden City, N.Y.: Doubleday & Company, 1973.

GRAYSON, M. J., and SHEPARD, T. R., JR. *The Disaster Lobby. Prophets of Ecological Doom and Other Absurdities.* Chicago: Follett, 1973.

GREEN, C. W. *Eli Whitney and the Birth of American Technology.* Boston: Little, Brown & Co., 1956.

GREENBERG, D. S. *The Politics of Pure Science.* New York: New American Library, 1968.

GROSS, BERTRAM. *Space-Time and Post-Industrial Society.* Washington, D.C.: Comparative Administration Group, American Society for Public Administration, 1966.

HABERMAS, JURGEN. *Toward a Rational Society: Student Protest, Science, and Politics.* Boston: Beacon Press, 1970.

HACKER, ANDREW. *The End of the American Era,* New York: Atheneum Publishers, 1968, 1970.

HAMILTON, DAVID. *Technology, Man and the Environment.* New York: Charles Scribner's Sons, 1973.

HAMILTON, MICHAEL, ed. *The New Genetics and the Future of Man.* Grand Rapids, Michigan: Eerdmans, 1972.

HAMMOND, PETER B., ed. *Cultural and Social Anthropology.* New York: Macmillan, 1964.

HANDLER, PHILIP, ed. *Biology and the Future of Man.* New York: Oxford University Press, 1970.

HARDIN, GARRETT. *Stalking the Wild Taboo.* Los Altos, California: William Kaufmann, 1973.

HARRINGTON, MICHAEL. *The Accidental Century.* New York: Macmillan Company, 1965.

HARRIS, LOUIS. *The Anguish of Change.* New York: W. W. Norton & Company, 1973.

HARVEY, M. L., et al. *Science and Technology as an Instrument of Soviet Policy.* Washington, D.C.: Center for Advanced International Studies, University of Miami, 1972.

HELLMAN, HAL. *Biology in the World of the Future.* New York: M. Evans and Company, 1971.

———. *Communications in the World of the Future.* 2nd ed. New York: M. Evans and Company, 1975.

———. *The City in the World of the Future.* New York: M. Evans and Company, 1970.

———. *Energy in the World of the Future.* New York: M. Evans and Company, 1973.

———. *Feeding the World of the Future.* New York: M. Evans and Company, 1972.

———. *Transportation in the World of the Future.* 2nd ed. New York: M. Evans and Company, 1974.

———, ed. *Epidemiological Aspects of Carcinogenesis.* Washington, D.C.: Interdisciplinary Communications Program, Smithsonian Institution, 1973.

———, ed. *Migration, Urbanization, Fertility.* Washington, D.C.: Interdisciplinary Communications Program, Smithsonian Institution, in press.

HELMER, OLAF. *Social Technology,* with contributions by Bernice Brown and Theodore Gordon. New York: Basic Books, 1966.

HIGBEE, EDWARD. *A Question of Priorities; New Strategies for our Urbanized World,* New York: William Morrow and Company, 1970.

HILL, JAMES J. *Highways of Progress.* New York: Doubleday, Page & Company, 1910.

HILLEGAS, M. R. *The Future as Nightmare: H. G. Wells and the Anti-Utopians.* New York: Oxford University Press, 1967.

HOFFER, ERIC. *First Things, Last Things.* New York: Harper & Row, Publishers, 1971.

———. *The Ordeal of Change.* New York: Harper & Row, 1952; rev. ed., 1963.

HOFSTADTER, RICHARD. *Anti-Intellectualism in American Life.* New York: Alfred A. Knopf, 1963.

HOLZER, HANS. *The Truth About Witchcraft.* Garden City, N.Y.: Doubleday & Company, 1969.

HOOVER, HELEN. *The Years of the Forest.* New York: Alfred A. Knopf, 1973.

Horizon. "Reason and Unreason," Spring 1970. (Eight articles on the use and non-use of reason: Past, present and future.)

HUBBARD, L. R. *Dianetics, The Modern Science of Mental Health.* New York: Paperback Library, 1968.

HUDSON, K. *Industrial Archaeology.* Chester Springs, Pennsylvania: Dufour Editions, 1964.

HUXLEY, ALDOUS. *Brave New World Revisited.* New York: Harper & Row, Publishers, 1958.

———. *Tomorrow and Tomorrow and Tomorrow, and Other Essays.* New York: Harper & Brothers, 1952, 1953, 1955, 1956.

ILLICH, IVAN. *Tools for Conviviality.* New York: Harper & Row, Publishers, 1973.

INKELES, A., and SMITH, D. H. *Becoming Modern. Individual Change in Six Developing Countries.* Cambridge, Mass.: Harvard University Press, 1974.

INNES, HAROLD A. *The Bias of Communication.* Toronto: University of Toronto Press, 1951.

JACOBS, JANE. *The Life and Death of Great American Cities.* New York: Random House, 1961.

JACOBY, N. H. *Corporate Power and Social Responsibility.* New York: Macmillan Company, 1973.

JAHODA, GUSTAV. *The Psychology of Superstition.* Baltimore: Penguin Books, 1969.

JAY, MARTIN. *The Dialectical Imagination. A History of the Frankfurt School and the Institute of Social Research, 1923–1950.* Boston: Little, Brown and Company, 1973.

JESSOPP, AUGUSTUS. *Arcady, For Better or Worse.* London: T. F. Unwin, 1892.

JEWKES, J.; SAWYERS, D.; and STILLERMAN, R. *The Sources of Invention.* New York: St. Martin's Press, 1958.

JOSEPH, S. M. *Children in Fear.* New York: Holt, Rinehart and Winston, 1974.

JOUVENAL, BERTRAND DE. *The Art of Conjecture.* New York: Basic Books, 1967.

JUENGER, FRIEDRICH GEORG. *The Failure of Technology.* Chicago: Henry Regnery Company, 1946; paperback, 1956.

JUNGK, ROBERT. *Tomorrow Is Already Here.* New York: Simon and Schuster, 1954.

KAHLER, ERICH. *The Tower and the Abyss.* New York: Viking Press, 1967.

KETTERER, DAVID. *New Worlds for Old: The Apocalyptic Imagination, Science Fiction, and American Literature.* Garden City, N.Y.: Doubleday & Company, 1974.

KIDD, BENJAMIN. *Social Evolution.* New York: Macmillan and Company, 1895.

KATEB, GEORGE. *Utopia and Its Enemies.* New York: Free Press paperback, 1963.

KOESTLER, ARTHUR. *The Ghost in the Machine.* New York: Macmillan Company, 1968.

KONE, E. H., and JORDAN, H. J., eds. *The Greatest Adventure: Basic Research That Shapes Our Lives.* New York: Rockefeller University Press, 1974.

KOSTELANETZ, RICHARD, ed. *The Edge of Adaptation. Man and the Emerging Society.* Englewood Cliffs, N.J.: Prentice-Hall, 1974.

——. *The New American Arts.* New York: Collier Books, 1967.

KRANZBERG, M., and DAVENPORT, W. H., eds. *Technology and Culture.* New York: Schocken Books, 1972.

KRANZBERG, M., and PURSELL, C. W., eds. *Technology in Western Civilization,* vols. 1 and 2. New York: Oxford University Press, 1967.

KUHNS, WILLIAM. *The Post-Industrial Prophets; Interpretations of Technology.* New York: Weybright & Talley, 1971.

LAING, R. D. *The Politics of Experience,* Baltimore: Penguin Books, 1967.

LAIRD, D. A., and LAIRD, E. C. *How to Get Along With Automation.* New York: McGraw-Hill, 1964.

LANDERS, R. R. *Man's Place in the Dybosphere.* Englewood Cliffs, N.J.: Prentice-Hall, 1966.

LAPP, R. E. *The New Priesthood.* New York: Harper & Row, Publishers, 1965.

——. *The Weapons Culture.* New York: W. W. Norton & Company, 1968.

LASLETT, PETER. *The World We Have Lost.* New York: Charles Scribner's Sons, 1966.

LEISS, WILLIAM. *The Domination of Nature.* New York: George Braziller, 1972.

LEONARD, G. *The Transformation.* New York: Delacorte Press, 1972.

LE ROY, LOUIS. *Of the Interchangeable Course or Varity of Things in the Whole World: And the Concurrence of Arms and Learning Through the First and Famousest Nations: From the Beginning of Civility and Memory of Man to This Present.* London: Charles Yetsweirt, 1594.

LEVY, ELIZABETH. *The People Lobby: The SST Story.* New York: Delacorte Press, 1973.

LEVY, LILLIAN, ed. *Space: Its Impact Upon Man and Society.* New York: W. W. Norton & Company, 1965.

LEWIS, A. O., JR. *Of Men and Machines.* New York: E. P. Dutton & Co., 1963.

LIVINGSTON, J. A. *One Cosmic Instant. Man's Fleeting Supremacy.* Boston: Houghton Mifflin, 1973.

LUNDBORG, L. B. *Future Without Shock.* New York: W. W. Norton & Company, 1974.

MADDOX, JOHN. *The Doomsday Syndrome.* New York: McGraw-Hill, 1972.

MANSFIELD, EDWIN. *Industrial Research and Technological Innovation.* New York: W. W. Norton, 1967.

MARCEL, GABRIEL. *Man Against Mass Society.* Chicago: Henry Regnery Company, 1952.

MARCUSE, HERBERT. *Counterrrevolution and Revolt.* Boston: Beacon Press, 1972.

———. *One-Dimensional Man.* Boston: Beacon Press, 1964.

MARSAK, LEONARD, ed., *The Rise of Science in Relation to Society.* New York: Macmillan Company, 1964.

MARX, LEO. *The Machine in the Garden.* New York: Oxford University Press, 1964.

MCCLUHAN, MARSHALL. *Understanding Media.* New York: McGraw-Hill, 1965.

MCPHEE, JOHN. *The Curve of Binding Energy.* New York: Farrar, Straus & Giroux, 1974.

MEAD, MARGARET, ed. *Cultural Patterns and Technical Change.* New York: Mentor Books, 1955.

MEADOWS, D. L., et al. *The Limits to Growth.* New York: Universe Books, 1972.

MEDFORD, DEREK. *Environmental Harassment or Technology Assessment?* New York: Elsevier, 1973.

MERTON, R. K. *The Sociology of Science.* Edited by N. W. Storer. Chicago: University of Chicago Press, 1973.

MESTHENE, E. G. *Technological Change; Its Impact on Man and Society.* Cambridge, Mass.: Harvard University Press, 1970.

METZ, L. D., and KLEIN, R. E. *Man and the Technological Society.* Englewood Cliffs, N.J.: Prentice-Hall, 1973.

MICHAEL, D. N. *Cybernation: The Silent Conquest.* Santa

Barbara, Calif.: Center for the Study of Democratic Institutions, 1962.

———. *The Unprepared Society, Planning for a Precarious Future.* New York: Basic Books, 1968.

MILLS, C. W. *The Power Elite.* New York: Oxford University Press, 1965.

MISHAN, E. J. *Technology and Growth: The Price We Pay.* New York: Praeger Publishers, 1969.

MITCHELL, J. G., and STALLINGS, C. L., eds. *Ecotactics: The Sierra Club Handbook for Environmental Activists.* New York: Pocket Books, 1970.

MONOD, JACQUES. *Chance and Necessity: An Essay on the Natural Philosophy of Modern Biology.* New York: Alfred A. Knopf, 1971.

MOORE, W. G., ed. *Technology and Social Change.* New York: Quadrangle Books, 1972.

MORISON, E. E. *From Know-How to Nowhere: The Development of American Technology.* New York: Basic Books, 1975.

MORSE, D., and WARNER, A. W., eds. *Technological Innovation and Society.* New York: Columbia University Press, 1966.

MOSS, THELMA. *The Probability of the Impossible: Scientific Discoveries and Explorations in the Psychic World.* New York: Hawthorne Books, 1974.

MULLER, H. J. *The Children of Frankenstein, A Primer on Modern Technology and Human Values.* Bloomington: Indiana University Press, 1970.

———. *Man's Future Birthright: Essays on Science and Humanity.* Albany: State University of N.Y. Press, 1973.

MUMFORD, LEWIS. *The Pentagon of Power; The Myth of the Machine.* Vol. 2, New York: Harcourt, Brace & Jovanovich, 1970.

———. *Technics and Civilization.* New York: Harcourt, Brace and World, 1963.

MURPHY, E. F. *Governing Nature.* New York: Quadrangle Books, 1968.

NADER, RALPH, et al., eds. *Whistle Blowing.* New York: Bantam Books, 1972.

NEARING, SCOTT. *Freedom: Promise and Menace.* Harborside, Maine: Social Science Institute, 1961.

NEEDLEMAN, JACOB. *The New Religions.* Garden City, N.Y.: Doubleday, & Company, 1970.

OLIVER, J. W. *History of American Technology.* New York: Ronald Press, 1966.

ORTEGA Y GASSET, JOSE. *The Revolt of the Masses.* New York: W. W. Norton & Company, 1932.

OSBORNE, J. W. *The Silent Revolution: The Industrial Revolution in England as a Source of Social Change.* New York: Charles Scribner's Sons, 1970.

OSTRANDER, G. M. *American Civilization in the First Machine Age, 1890–1940.* New York: Harper & Row, Publishers, 1970.

OVERSTREET, B. W. *Understanding Fear.* New York: Harper & Brothers, 1951.

PADDOCK, P., and PADDOCK, W. *Famine 1975!* Boston: Little, Brown and Company, 1967.

PAGLIARO, H. E., ed. *Irrationalism in the 18th Century, (Proceedings of the American Society for Eighteenth Century Studies, 1971).* Cleveland: University Press of Case Western Reserve, 1972.

PARKER, RICHARD. *The Myth of the Middle Class.* New York: Liveright Publishing Co., 1972.

PETERSEN, WM. J. *Those Curious New Cults.* New Canaan, Connecticut: Keats Publishing Inc., 1973.

PINES, MAYA. *The Brain Changers; Scientists and the New Mind Control.* New York: Harcourt, Brace & Jovanovich, 1973.

PLATT, J. R. *The Step to Man.* New York: John Wiley & Sons, 1966.

PRICE, D. K. *The Scientific Estate.* Cambridge, Mass.: Harvard University Press, 1965.

PRIESTLEY, J. B. *Victoria's Heyday,* New York: Harper & Row, Publishers, 1972.

PURCELL, E. A., JR. *The Crisis of Democratic Theory. Scientific Naturalism and the Problem of Value.* Lexington: University Press of Kentucky, 1973.

QUEFFELAC, HENRI. *Technology and Religion.* New York: Hawthorne Books, 1964.

REICH, C. A. *The Greening of America.* New York: Random House, 1970.

REVEL, J. F. *Without Marx or Jesus.* Garden City, N.Y.: Doubleday & Company, 1971.

RIESMAN, DAVID. *Abundance for What?* Garden City, N.Y.: Doubleday & Company, 1964.

ROCKS, L., and RUNYON, R. P. *The Energy Crisis.* New York: Crown Publishers, 1972.

RORVIK, D. M. *As Man Becomes Machine: The Evolution of the Cyborg.* Garden City, N.Y.: Doubleday & Company, 1971.

ROSENBERG, NATHAN. *Technology and American Economic Growth.* New York: Harper & Row, Publishers, 1972.

ROSLANSKY, J. D., ed. *Genetics and the Future of Man.* New York: Appleton-Century-Crofts, 1966.

ROSZAK, T. *The Making of a Counter-Culture.* Garden City, N.Y.: Doubleday & Company, 1969.

————. *Where the Wasteland Ends.* Garden City, N.Y.: Doubleday & Company, 1972.

ROTHSCHILD, EMMA. *Paradise Lost: The Decline of the Auto-Industrial Age.* New York: Random House, 1973.

ROWLEY, PETER. *New Gods in America.* New York: David McKay, 1971.

SCHON, D. A. *Technology and Change.* New York: Delacorte Press, 1967.

SCHWARTZ, E. S. *Overskill.* New York: Quadrangle Books, 1971.

SCHWARTZ, HARRY. *The Case for American Medicine.* New York: David McKay, 1972.

SCHWITZGEBEL, R. L., and SCHWITZGEBEL, R. K., eds. *Psychotech-*

nology: Electronic Control of Mind and Behavior. New York: Holt, Rinehart and Winston, 1973.

SCOTT, RACHEL. *Muscle and Blood, the Massive, Hidden Agony of Industrial Slaughter in America.* New York: E. P. Dutton & Co., 1974.

SELIGMAN, B. B. *Most Notorious Victory.* New York: Free Press, 1966.

SERLING, R. J. *Loud and Clear.* New York: Dell Books, 1970.

SERVAN-SCHREIBER, J. J. *The American Challenge.* New York: Atheneum Publishers, 1968.

SHEPARD, PAUL. *The Tender Carnivore and the Sacred Game.* New York: Charles Scribner's Sons, 1973.

SHILS, EDWARD. *The Intellectuals and the Powers and Other Essays.* Chicago: University of Chicago Press, 1972.

SILBERMAN, C. E. *The Myth of Automation.* New York: Harper & Row, Publishers, 1966.

SKINNER, B. F. *About Behaviorism.* New York: Alfred A. Knopf, 1974.

SLATER, PHILIP. *Earthwalk.* Garden City, N.Y.: Doubleday, 1974.

SLOMICH, S. J. *The American Nightmare.* New York: Macmillan Company, 1971.

SNOW, C. P. *The State of Siege.* New York: Charles Scribner's Sons, 1969.

———. *The Two Cultures and the Scientific Revolution.* Cambridge: Cambridge University Press, 1961.

SOLERI, PAOLO. *Arcology: the City in the Image of Man.* Cambridge, Mass. MIT Press, 1969.

SPENCER, CORNELIA. *Keeping Ahead of Machines; the Human Side of the Automation Revolution.* New York: John Day Company, 1965.

SPICER, E. H., ed. *Human Problems in Technological Change.* New York: Russell Sage Foundation, 1952.

STAMBLER, IRWIN. *Automobiles of the Future.* New York: Putnam, 1966.

STOBER, G. J., and SCHUMACHER, D., eds. *Technology Assessment and Quality of Life.* Amsterdam: Elsevier, 1973.

SUSSMAN, H. L. *Victorians and the Machine*. Cambridge, Mass.: Harvard University Press, 1968.

SYPHER, WYLIE. *Literature and Technology. The Alien Vision*. New York: Random House, 1968.

TAVISS, IRENE. *Our Tool-Making Society*. Englewood Cliffs, N.J.: Prentice-Hall paperback, 1972.

TAYLOR, G. R. *The Biological Time Bomb*. Cleveland: World Publishing Company, 1968.

TEILHARD DE CHARDIN, PIERRE, *The Future of Man*. London: Fontana Books, 1964; New York: Harper & Row, 1969.

———. *The Phenomenon of Man*. New York: Harper & Row, Publishers, 1959.

THEOBALD, R. A. *The Challenge of Abundance*. New York: Mentor Books, 1962.

THEOBALD, R. A., ed. *Dialogue on Technology*. Indianapolis: Bobbs-Merrill Company, 1967.

THRING, M. W., and BLAKE, A. *Man, Machines and Tomorrow*. Boston: Routledge & Kegan Paul, 1973.

DeTOCQUEVILLE, ALEXIS. *Democracy in America*. 2 vols. New York: Alfred A. Knopf, 1944.

TOFFLER, ALVIN. *Future Shock*. New York: Random House, 1970.

TOURAINE, ALAIN. *The Post-Industrial Society: Tomorrow's Social History; the Programmed Society, the New Classes, the Managers and the Managed*. New York: Random House, 1967.

VALENSTEIN, ELLIOT S. *Brain Control. A Critical Examination of Brain Stimulation and Psychosurgery*. New York: Wiley-Interscience, 1973.

VAN LEEWVEN, A. T. *Prophesy in a Technocratic Era*. New York: Charles Scribner's Sons, 1968.

VERRETT, J. and CARPER, J. *Eating May Be Hazardous to Your Health: How Your Government Fails to Protect You from the Dangers in Your Food*. New York: Simon and Schuster, 1974.

VON BERTALANFFY, LUDWIG. *Robots, Men and Minds*. New York: George Braziller, 1967.

VON DANIKEN, E. *In Search of Ancient Gods, My Pictorial Evidence for the Impossible*. New York: G. P. Putnam's Sons, 1973, 1974.

WAGAR, W. W., ed. *The City of Man*. Rev. ed. Baltimore: Penguin Books, 1968.

WALKER, C. R. *Modern Technology and Civilization: An Introduction to Human Problems in the Machine Age*. McGraw-Hill, 1962.

WATSON, LYALL. *Supernature*. Garden City, N.Y.: Doubleday & Company, 1973.

WEEKES, CLAIRE. *Peace From Nervous Suffering*. New York: Hawthorne Books, 1972.

WEINBERG, ALVIN. *Reflections on Big Science*. Cambridge, Mass.: MIT Press, 1967.

WEST, J. A., and TOONDER, J. G. *The Case for Astrology*. New York: Coward, McCann & Geohegan, 1970.

WESTIN, A. F., and BAKER, M. A. *Databanks in a Free Society. Computers, Record-Keeping and Privacy*. New York: Quadrangle Books, 1973.

WHEELIS, ALLEN. *The End of the Modern Age*. New York: Basic Books, 1972.

WHITE, MORTON. *Science and Sentiment in America*. New York: Oxford University Press, 1972.

WHITE, M., and WHITE, L. *The Intellectual Versus the City*. New York: Mentor Books, 1962.

WHITEHEAD, ALFRED NORTH. *Science and the Modern World*. New York: Macmillan, 1925. Mentor paperback, 1948.

———. *The Human Use of Human Beings*. Boston: Houghton Mifflin, 1950.

WIENER, NORBERT. *God and Golem, Inc*. Cambridge, Mass.: MIT Press, 1964.

WILKINSON, JOHN, et al. *Technology and Human Values*. Santa Barbara, Calif.: Center for the Study of Democratic Institutions, 1966.

WILLRICH, MASON, and TAYLOR, THEODORE B. *Nuclear Theft: Risks and Safeguards.* Cambridge, Mass.: Ballinger Publishing Co., 1974.

WOLFE, TOM. *The Electric Kool-Aid Acid Test.* New York: Farrar, Straus & Giroux, 1968.

————. *The Kandy-Kolored Tangerine-Flake Streamline Baby.* New York: Farrar, Straus & Giroux, 1965.

WOOD, ANTHONY. *Nineteenth Century Britain, 1815–1914.* London: Longmans, Green & Co., 1960.

WOOLDRIGE, D. E. *Mechanical Man: The Physical Basis of Intelligent Life.* New York: McGraw-Hill, 1968.

YOUNG, L. B. *Power Over People.* New York: Oxford University Press, 1973.

YOUNG, MICHAEL. *The Rise of the Meritocracy 1870–2033.* London: Thames and Hudson, 1958. Penguin paperback, 1961.

YOUNGBLOOD, GENE. *Expanded Cinema.* New York: Dutton, 1970.

ZARETSKY, I. I., and LEONE, M. P. *Religious Movements in Contemporary America,* Princeton: Princeton University Press, 1975.

Articles:

AARONSON, STEVE. "Pictures of an Unknown Aura." *The Sciences,* January/February 1974.

ABELSON, P. H. "Federal Support of Graduate Education." *Science,* 3 March 1972.

————. "Unrealistic Demands on Science and Medicine." *Science,* 4 June 1971.

ADLER, ALFRED. "Science and Evil." *Atlantic,* February 1972.

ADLER, NATHAN. "The Antinomian Personality: The Hippie Character Type." *Psychiatry,* November 1968.

AIGRAIN, PIERRE. "Trends in Research Funding." *Physics Today,* November 1974.

ALEXANDER, TOM. "The Hysteria About Food Additives." *Fortune,* March 1972.

――――. "Ominous Changes in the World's Weather." *Fortune*, February 1974.

ALONSO, WILLIAM. "Urban Zero Population Growth." *Daedalus*, Fall 1973.

ANDERSON, ALAN, JR. "The Hidden Plague." *New York Times Magazine*, 27 October 1974.

APODACA, ANACLETO. "Corn and Custom: The Introduction of Hybrid Corn to Spanish American Farmers in New Mexico." In E. H. Spicer, ed., *Human Problems in Technological Change*. New York: Russell Sage Foundation, 1952, p. 35.

AREHART-TREICHEL, JOAN. "Biological Controls for Insects Are Here . . ." *Science News*, 21 July 1973.

――――. "Green Revolution: Phase 2." *Science News*, 21 July 1933.

ASTURIAS, MIGUEL ANGEL. "A Dream of Waters Glittering With Stars." *Impact of Science on Society*, October–December 1969.

ATKINS, HARRY. "That New College Spirit." *The Sciences*, January/February 1972.

ATTAH, E. B. "Racial Aspects of Zero Population Growth." *Science*, 15 June 1973.

AVORN, JERRY. "The Varieties of Postpsychedelic Experience," an interview with Robert Masters and Jean Houston. *Intellectual Digest*, March 1973.

BARAM, M. S. "Social Control of Science and Technology." *Science*, 7 May 1971.

――――. "Technology Assessment and Social Control." *Science*, 4 May 1973.

BAZELL, R. J. "NSF: Is Applied Research at the Take-Off Point?" *Science*, 25 June 1971.

BELL, DANIEL. "The End of Scarcity?" *Saturday Review of Society*, May 1973.

――――. "Notes on the Post-Industrial Society." *Public Interest*, Winter 1967 and Spring 1967.

――――. "Meritocracy and Equality." *Public Interest*, Fall 1972.

BENN, A. W. "Technical Power and People: The Impact of Technology on the Structure of Government." *Bulletin of the Atomic Scientists,* December 1971.

BERG, C. A. "Energy Conservation Through Effective Utilization." *Science,* 13 July 1973.

BERGER, BENNETT. "Hippie Morality—More Old Than New." *Trans-Action,* December 1967.

BJORK, L. E. "An Experiment in Work Satisfaction." *Scientific American,* March 1975.

BLISS, W. L. "In the Wake of the Wheel: Introduction of the Wagon to the Papago Indians of Southern Arizona." In E. H. Spicer, ed. *Human Problems in Technological Change.* New York: Russell Sage Foundation, 1952.

BOETH, R. "Assault on Privacy." *Newsweek,* 27 July 1970.

BOFFEY, P. M. "Science Policy: An Insider's View of LBJ, DuBridge, and the Budget." *Science,* 5 March 1971.

———. "Technology and World Trade: Is There Cause for Alarm?" *Science,* 2 April 1971.

BORETSKY, MICHAEL. "Trends in U. S. Technology: A Political Economist's View." *American Scientist,* January-February 1975.

BOORSTIN, D. J. "A Case of Hypochondria," in "The Spirit of '70: Six Historians Reflect on What Ails the American Spirit." *Newsweek,* 6 July 1970.

BORKENAU, FRANZ. "Will Technology Destroy Civilization?" *Commentary,* January 1951.

BORN, MAX. "Blessings and Evils of Space Travel." *Bulletin of the Atomic Scientists,* October 1966.

BOULDING, K. E. "The Scientific Revelation." *Bulletin of the Atomic Scientists,* September 1970.

BOWERS, RAYMOND. "Some Views on Physics and Society." *Physics Today,* November–December 1970.

BOWERS, R., and FREY, J. "Technology Assessment and Microwave Diodes." *Scientific American,* February 1972.

BRANSCOMB, L. M. "Managing Technology By Performance." *Science,* 21 January 1972.

————. "Taming Technology." *Science,* 12 March 1971.

————. "Why People Fear Technology." *The Futurist,* December 1971.

BRINCKLOE, W. D. "Automation and Self-Hypnosis." *Public Administration Review,* September 1966.

BRODE, W. R. "Manpower in Science and Engineering, Based on a Saturation Model." *Science,* 16 July 1971.

BRONOWSKI, J. "The Disestablishment of Science." *Encounter,* July 1971.

BRONSON, G. W. "The Development of Fear in Man and Other Animals." *Child Development,* June 1968.

BROOKS, HARVEY. "Are Scientists Obsolete?" *Science,* 8 November 1974.

————. "Can Science Survive in the Modern Age?" *Science,* 10 October 1971.

————. "Physics and the Polity." *Science,* 26 April 1968.

————. "The Technology of Zero Growth." *Daedalus,* Fall 1973.

BROOKS, H., and BOWERS, R. "The Assessment of Technology." *Scientific American,* February 1970.

BROWN, G. E., JR. "Physics and Social Change." *Physics Today,* October 1971.

BROWN, HARRISON. "Science, Technology, and the Developing Countries." *Bulletin of the Atomic Scientists,* June 1971.

BROWN, LESTER. "Rich Countries and Poor in a Finite, Interdependent World." *Daedalus,* Fall 1973.

BROWN, N. O. "Apocalypse: The Place of Mystery in the Life of the Mind." *Harpers,* May 1961.

BRZEZINSKI, ZBIGNIEW. "America in the Technetronic Age." *Encounter,* January 1968.

————. "Toward a Technetronic Society." *Current,* February 1968.

BURCK, GILBERT. "There'll Be Less Leisure Than You Think." *Fortune,* March 1970.

BURSTEIN, S. M. "Science, Abraham & Ecology." *Intellectual Digest,* October 1972.

BUSH, VANNEVAR. "Dictation to Science by Laymen." *Science,* October 1971.

BUTTERFIELD, HERBERT. "The Scientific Revolution." *Scientific American,* September, 1960.

BYLINSKY, GENE. "New Clues to the Causes of Violence." *Fortune,* January 1973.

CADWALADER, GEORGE. "Freedom for Science in the Oceans." *Science,* 5 October 1973.

CALDER, NIGEL. "Food From Gas, Oil." *New Scientist,* 23 November 1967.

———."Tomorrow's Politics: The Control and Use of Technology." *The Nation,* 4 January 1965.

CALDER, RITCHIE. "Speed of Change." *Bulletin of the Atomic Scientists,* December 1965.

CALDWELL, L. C. "Managing the Scientific Super-Culture: The Task of Educational Preparation." *Public Administration Review,* June 1967.

CARLYLE, THOMAS. "Signs of the Times" *(Edinburgh Review, 1829).* In Thomas Carlyle, *Scottish and Other Miscellanies.* New York: E. P. Dutton & Co.

CARROLL, J. D. "Participatory Technology." *Science,* 19 February 1971.

———. "Science and the City: the Question of Authority." *Science,* 28 February 1969.

CARTER, A. P. "Economics of Technological Change." *Scientific American,* April 1966.

CARTER, LUTHER J. "Weather Modification: Colorado Heeds Voters in Valley Dispute." *Science,* 29 June 1973.

CARTTEN, A. M. "Scientific Manpower for 1970–1985." *Science,* 9 April 1971.

CASPER, B. M. "Physicists and Public Policy: The 'Forum' and the APS." *Physics Today,* May 1974.

CHARLES, K. J. "The Third World's Struggle to Use People Power." *The Futurist,* December 1974.

CHOROVER, S. L. "Big Brother and Psychotechnology." *Psychology Today,* October 1973.

CLARK, WILSON. "It Takes Energy to Get Energy; the Law of

Diminishing Returns Is In Effect," *Smithsonian*, December 1974.

CLARKE, ROBIN. "Technology for an Alternative Society." *New Scientist*, January 11, 1973.

COATES, J. F. "Technology Assessment: The Benefits . . . The Costs . . . The Consequences." *The Futurist*, December 1971.

COATES, V. T. "Technology in the Balance." *The Futurist*, April 1973.

COHEN, B. L. "Learning to Live With Radiation." *Science Digest*, April 1975.

COHN, VICTOR. "Anti-Science and the Energy Crisis." *Technology Review*, February 1974.

COLE, J. R., and COLE, S. "The Ortega Hypothesis." *Science*, 27 October 1973.

COLE, LAMONT C. "Can the World Be Saved?" *New York Times Magazine*, 31 March 1968.

COLLIGAN, DOUGLAS. "How to Recycle a Cow Burp. Or Some Imaginative Energy Choices for the Future." *Science Digest*, October 1974.

COUDERC, PAUL. "An Antidote for Anti-Science." *Impact*, April–June 1971.

COULOMB, JEAN. "Good Use of Scientists." *Bulletin of the Atomic Scientists*, January 1971.

COX, HARVEY. "God and the Hippies." *Playboy*, January 1968.

CRANE, THEODORE. "The Squalor That Was Rome." *Natural History*, May 1973.

DAVIDSON, F. P. "The Case for Institutional Assessment." *Technology Review*, December 1971.

DAVIS, F. "Why All of Us May Be Flower Children Some Day." *Trans-Action*, December 1967.

DEAN, N. W. "Modeling PhD Production 1960–2000." *Physics Today*, May 1973.

DeMAUSE, LLOYD. "Our Forebears Made Childhood a Nightmare." *Psychology Today*, April 1975.

DEUTSCH, K. W., et al. "Conditions Favoring Major Advances in Social Science." *Science*, 5 February 1971.

DIXON, BERNARD. "Scientists and Social Ferment." *New Scientist*, 5 July 1973.

DOUGLAS, J. H. "Climate Change: Chilling Possibilities." *Science News*, March 1, 1975.

———. "Science Education: The New Humanity." *Science News*, 24 March 1973.

———. "Turning from Science: Shortages [in Manpower] Ahead." *Science News*, 24 March 1973.

DRUCKER, D. C. "The Engineer in the Establishment." *Bulletin of the Atomic Scientists*, December 1971.

DUBOS, RENE. "Health and Environment." *American Lung Association Bulletin*, September 1973.

———. "Science and Man's Nature." *Daedalus*, Winter 1965.

———. "Scientists Alone Can't Do the Job." *Saturday Review*, 2 December 1967.

———. "A Social Design for Science." *Science*, 14 November 1969.

DYSON, F. J. "The Hidden Cost of Saying No!" *Bulletin of the Atomic Scientists*, June 1975.

EBERHART, JONATHAN. "Earthquake War Is Go." *Science Digest*, 29 September 1973.

EDSON, LEE. "The Psyche and the Surgeon." *New York Times Magazine*, 30 September 1973.

EISENBUD, MERRIL. "Environmental Protection in the City of New York." *Science*, 13 November 1970.

EMMONS, H. W. "Fire and Fire Protection." *Scientific American*, July 1974.

ESFANDIARY, F. M. "The Mystical West Puzzles the Practical East." *New York Times Magazine*, 6 February 1967.

ETZIONI, A., and REMP, R. "Technological 'Shortcuts' to Social Change." *Science*, 7 January 1972.

FEIT, H. A. "Twilight of the Cree Hunting Nation." *Natural History*, August/September 1973.

FERRY, W. H. "Must We Rewrite the Constitution to Control Technology?" *Saturday Review*, 2 March 1968.

FISHER, JOHN. "Why Our Scientists Are About to Be Dragged, Moaning, Into Politics." *Harpers*, September 1966.

FLING, KAREN. "Funding for Academic Science Cut." *BioScience*, 15 May 1971.

FOECKE, H. A. "Engineering in the Humanistic Tradition," *Impact*, April/June 1970.

FOY, NANCY. "A Tale of Three Factories." *New Scientist*, 10 October 1974.

FRANK, PHILIP. "Philosophical Uses of Science." *Bulletin of the Atomic Scientists*, April 1957.

FRANKEL, CHARLES. "The Nature and Sources of Irrationalism." *Science*, 1 June 1973.

FREESE, ARTHUR. "Ominous Miracle Drugs: Antibiotics That Kill." *Science Digest*, January 1975.

FRIEDENBERG, E. Z. "The Hidden Costs of Opportunity." *Atlantic Monthly*, February 1969.

———. "The Hostile Delusions of the Working Class." *Harpers*, June 1973.

Fuller, R. B. "The Age of Astro-Architecture," *Saturday Review*, 13 July 1968.

———. "Goddesses of the Twenty-First Century," *Saturday Review*, 2 March 1968.

———. "Vision 65." *The American Scholar*, Spring 1966.

GAFFRON, HANS. "Resistance to Knowledge," *Annual Review of Plant Physiology*, 20, (1969). Expanded and republished in 1970 as Occasional Papers 2 of the Salk Institute, San Diego, California.

GALSTON, A. W. "Education of a Scientific Innocent." *Natural History*, June/July 1971.

———. "Hard Times for American Science," *Natural History*, August/September 1973.

GANZ, A., and O'BRIEN, T. "New Directions for Our Cities in the Seventies." *Technology Review*, June 1974.

GANZ, HERBERT. "The New Egalitarianism." *Saturday Review*, 6 May 1972.

GAY, PETER. "The Enlightenment." *Horizon*, Spring 1970.

GEE, SHERMAN. "Foreign Technology and the United States Economy." *Science,* 21 February 1975.

GILLETTE, ROBERT. "Science in Mexico (I): The Revolution Seeks a New Ally." *Science,* 15 June 1973.

———. "Latin America: Is Imported Technology Too Expensive?" *Science,* 6 July 1973.

GLASS, BENTLEY. "Science: Endless Horizon or Golden Age." *Science,* 8 January 1971.

GOLDMAN, M. I. "The Convergence of Environmental Disruption." *Science,* 2 October 1970.

GOMER, ROBERT. "The Tyranny of Progress." *Bulletin of the Atomic Scientists,* February 1968.

GORDON, T. J. "The Effects of Technology on Man's Environment." *Architectural Design,* February 1967.

GOULDNER, H. P. "Children of the Laboratory." *Trans-Action,* April 1967.

GOULET, DENIS. "The Paradox of Technology Transfer." *Bulletin of the Atomic Scientists,* June 1975.

GRABINER, J. V., and MILLER, P. D. "Effects of the Scopes Trial. Was It a Victory for the Evolutionists?" *Science,* 6 September 1974.

GREEN, H. P. "The New Technological Era: A View From the Law." *Bulletin of the Atomic Scientists,* November 1967.

GREENBERG, D. S., JR. "The Myth of the Scientific Elite." *Public Interest,* Fall 1965.

GREENBERGER, M., et al. "Computer and Information Networks." *Science,* 5 October 1973.

HAKE, BARRY. "Values, Technology and the Future." *Futures Conditional,* June 1973.

HALL, D. T., and LAWLER, E. E. III. "Job Pressures and Research Performance." *American Scientist,* January/February 1971.

HANDLER, PHILIP. "The Federal Government and the Scientific Community." *Science,* 15 January 1971.

HANDLIN, OSCAR. "Science and Technology in Popular Culture." *Daedalus,* Winter 1965.

HANLON, JOSEPH. "Doctors Diagnose Data Distress." *New Scientist*, 5 July 1973.

HARDIN, E. "The Reactions of Employees to Office Automation." *Monthly Labor Review*, September 1960.

HARDIN, GARRETT. "The Tragedy of the Commons." *Science*, 25 June 1968.

HARRIS, MARVIN. "Potlatch Politics and Kings' Castles." *Natural History*, May 1974.

HARTLEY, W., and HARTLEY, E. "Phobias. Coping With Irrational Fears." *Science Digest*, September 1974.

HAYWARD, J. F. "The Uses of Myth in an Age of Science." *Zygon*, June 1968.

HELLMAN, HAL. "The Development of Inertial Navigation." *Navigation*, Summer 1962.

———. "The Future of the Automobile." *Future*, in press.

———. "The Story Behind Those Meatless Meats." *Popular Science*, 1972.

HENDERSON, HAZEL. "The Decline of Jonesism." *The Futurist*, October 1974.

HERMAN, E. S. "The Income Counter-Revolution." *Commonwealth*, 3 January 1975.

HIRSCHHORN, KURT. "On Re-Doing Man." *Commonwealth*, 17 May 1968.

HODGESON, J. A., et al. "Air Pollution Monitoring by Advanced Spectroscopic Techniques." *Science*, 19 October 1973.

HOHENEMSER, CHRISTOPH. "Science for the Concerned." *Environment*, August 1973.

HOLMBERG, A. R. "The Wells That Failed: An Attempt to Establish a Stable Water Supply in the Virn Valley, Peru." In E. H. Spicer, ed., *Human Problems in Technological Change.* (New York: Russell Sage Foundation), 1952.

HUDDLE, F. P. "The Social Management of Technological Consequences." *The Futurist*, February 1972.

HUXLEY, ALDOUS. "Brave New World Revisited." *Esquire*, October 1973.

IKLE, F. C. "Can Social Predictions Be Evaluated?" *Daedalus,* Summer 1967.

INHABER, H. "Environmental Quality: Outline for a National Index for Canada." *Science,* 29 November 1974.

INKELES, ALEX. "Six-Country Study of the Effects of Modernization." Chapter 5 in *International Collaboration in Mental Health.* Edited by B. S. Brown and G. F. Torrey. Washington, D.C.: National Institute of Mental Health, 1973.

INMAN, D. L., and BRUSH, B. M. "The Coastal Challenge." *Science,* 6 July 1973.

IYENGAR, M. S. "Can We Transform Into a Post-Industrial Society?" In *The Futurists.* Edited by A. Toffler. New York: Random House, 1972.

JAMES, R. D. "Measuring the Quality of Life." *The Wall Street Journal,* 18 May 1972.

JOHNSON, W. R. "Should the Poor Buy No Growth?" *Daedalus,* Fall 1973.

JOHNSTON, D. W. "Decline of DDT Residues in Migratory Song Birds." *Science,* 29 November 1974.

JONES, M. V. "The Methodology of Technology Assessment." *The Futurist,* February 1972.

JOUVENEL, BERTRAND dE. "The Political Consequences of the Rise of Science." *Bulletin of the Atomic Scientists,* December 1963.

KAHN, H., and WIENER, A. J. "Faustian Powers and Human Choices . . ." In W. R. Ewald, Jr., ed. *Environment and Change* . . . (Bloomington: Indiana University Press), 1968.

KALVIN, HARRY, JR. "The Problems of Privacy in the Year 2000." *Daedalus,* Winter 1967.

KAPLAN, MARTIN. "Science's Role in the World Health Organization." *Science,* 8 June 1973.

KECSKEMETI, PAUL. "Totalitarian Communications as a Means of Control." *Public Opinion Quarterly* 14 (1950), 224–234.

KELLY, C. F. "Mechanical Harvesting." *Scientific American,* August 1967.

KENDRICK, J. W. "Productivity: Can We Make Ours Grow Faster?" *Context* (DuPont), no. 3 (1973).

KENISTON, KENNETH. "You Have to Grow Up in Scarsdale to Know How Bad Things Really Are," *New York Times Magazine,* 27 April 1969.

———. "Youth, Change and Violence." *American Scholar,* Spring 1968.

KENNEDY E. M. "Statement on Re-employing Defense Scientists and Engineers." *Bulletin of the Atomic Scientists,* March 1971.

KENNER, HUGH. "Bucky Fuller and the Final Exam." *New York Times Magazine,* 6 July 1975.

KENWARD, MICHAEL. "Technology Transferred," *New Scientist,* 5 July 1973.

KLEIN, RUDOLPH. "Growth and Its Enemies." *Commentary,* June 1972.

———. "The Trouble With a Zero-Growth World." *New York Times Magazine,* 2 June 1974.

KOLSTAD, G. A. "What Future for Medium-Energy Physics?" *Physics Today,* February 1972.

KRAMER, EUGENE. "Energy Conservation and Waste Recycling, Taking Advantage of Urban Congestion." *Science and Public Affairs,* April 1973.

KRIEGER, DAVID. "Terrorists and Nuclear Technology," *Bulletin of the Atomic Scientists,* June 1975.

KRIEGER, M. H. "What's Wrong With Plastic Trees?" *Science,* 2 February 1973.

KRISTOL, IRVING. "It's Not Such a Bad Crisis to Live In," *New York Times Magazine,* 12 January 1967.

KRUTCH, J. W. "Is Life Just a Chemical Reaction?" *Saturday Review,* 4 May 1968.

LADD, E. C., JR., and LIPSET, S. M. "Politics of Academic Natural Scientists and Engineers." *Science,* 9 June 1972.

LAMBERT, DARWIN. "Personal Testimony on the Standard of Living." *National Parks,* June 1971.

LANSFORD, HENRY. "Weather Modification: The Public Will Decide." *Bulletin of the American Meteorological Society,* July 1973.

LEACH, GERALD. "Technophobia on the Left: Are British Intellectuals Anti-Science?" *New Statesman,* 27 August 1965.

LEAR, JOHN. "Policing the Consequences of Science," *Saturday Review,* 2 December 1967.

LEAR, JOHN, ed. "Science, Technology and the Law," *Saturday Review,* 3 August 1968.

LEONARD, EUGENE, et al. "MINERVA: A Participatory Technology System." *Bulletin of the Atomic Scientists,* November 1971.

LERNER, MAX. "Climate of Violence." *Playboy,* June 1967.

LESSING, LAWRENCE. "The Senseless War on Science." *Fortune,* March 1971.

LINCOLN, G. A. "Energy Conservation." Science, 13 April 1973.

LONG, F. A. "President Nixon's 1973 Reorganization Plan No. 1; Where Do Science and Technology Go Now?" *Science and Public Affairs,* May 1973.

LOPREATO, JOSEPH. "How Would You Like to Be a Peasant?" *Human Organization,* Winter 1965.

LOVE, SAM. "The Overconnected Society." *The Futurist,* December 1974.

LOVELL, BERNARD. "Serendipity in Science." *Intellectual Digest,* July 1973.

LOW, IAN. "A Future for the Scientist," *New Scientist,* 3 May 1973.

LOWINGER, PAUL. "Psychosurgery." *New Republic,* 13 April 1974.

LUKACS, JOHN. "So What Else Is New?" *New York Times Magazine,* 9 February 1975.

LUKASIEWICZ, J. "The Ignorance Explosion: A Contribution to the Study of Confrontation of Man With the Complexity of Science-Based Society and Environment." *Transactions of the New York Academy of Sciences,* May 1972.

MANSFIELD, E. "Contribution of R & D to Economic Growth in the United States." *Science*, 4 February 1972.

MARCUSE, HERBERT. "The Responsibility of Science," in: *The Responsibility of Power*. Edited by L. Krieger and F. Stern. Garden City, N.Y.: Doubleday & Company, 1967.

―――. "Some Social Implications of Modern Technology," *Studies in Philosophy and Social Science 9* (1941), 414–439.

MARTINO, J. P. "Adopting New Ideas." *The Futurist*, April 1974.

―――. "Can Computers Forecast Future Technological Developments?" *The Futurist*, August 1973.

―――. "The Pace of Technological Change." *The Futurist*, April 1972.

MARTY, M. E. "A Humanist's View of Space Research." *Chicago Today*, Autumn 1966.

MCCLINTOCK, ROBERT. "Machines and Vitalists: Reflections on the Ideology of Cybernetics," *American Scholar*, Spring 1966.

MCDERMOTT, JOHN. "Intellectuals and Technology." *New York Review of Books*, July 31, 1969.

MCELROY, W. D. "The Role of Fundamental Research in an Advanced Society." *American Scientist*, May/June 1971.

MCKEAN, R. N. "Growth vs No Growth: An Evaluation," *Daedalus*, Fall 1973.

MCKELVEY, JOHN. "Man vs. Fly." *R. F. Illustrated*, (Rockefeller Foundation) June 1973.

MCLUHAN, MARSHALL. Interview in *Playboy*, March 1969.

MCPHEE, JOHN. "Firewood," *New Yorker*, March 25, 1974.

MEDAWAR, P. B. "Man: The Technological Animal." *Intellectual Digest*, August 1973.

―――. "Science and the Sanctity of Life." *Encounter*, June 1966.

MEEHAN, THOMAS.. "The Flight From Reason." *Horizon*, Spring 1970.

MELMAN, SEYMOUR. "After the Military-Industrial Complex?" *Bulletin of the Atomic Scientists*, March 1971.

MELVILLE, KEITH. "A Measure of Contentment." *The Sciences*, December 1973.

MENCHER, A. G. "Management by Government: Science and Technology in Britain." *Bulletin of the Atomic Scientists,* May 1968.

———. "On the Social Deployment of Science." *Bulletin of the Atomic Scientists,* December 1971.

MESTHENE, E. G. "How Technology Will Shape the Future." *Science,* 12 July 1968.

———. "The Impacts of Science upon Public Policy." *Public Administration Review,* June 1967.

METZ, W. D. "Ocean Temperature Gradients: Solar Power From The Sea." *Science,* 22 June 1973.

MEZAN, PETER. "After Freud and Jung, Now Comes R. D. Laing." *Esquire,* January 1972.

MILTON, J. P. "Communities That Seek Peace With Nature." *The Futurist,* December 1974.

MITFORD, JESSICA. "The Torture Cure, Winning Criminal Hearts and Minds with Drugs, Scalpels and Sensory Deprivation." *Harpers,* August 1973.

MITZMAN, ARTHUR. "Anti-Progress: A Study in the Romantic Roots of German Sociology." *Social Research,* Spring 1966.

MORAMARCO, SHEILA. "Growing Plants with Seawater." *Science News,* 22 June 1974.

MORGENTHAU, H. J. "Modern Science and Political Power." *Columbia Law Review,* December 1964.

MOULIN, LEO. "The Nobel Prizes for the Sciences for 1901–1950: An Essay in Sociological Analysis." *British Journal of Sociology* 6 (1955), 246–263.

MOYNIHAN, D. P. "Nirvana." *American Scholar,* Autumn 1967.

MUMFORD, LEWIS. "Utopia, The City and the Machine." *Daedalus,* Spring 1965.

VAN NEUMANN, JOHN. "Can we Survive Technology?" *Fortune,* June 1955.

"Non-Scientists Dissect Science." *Impact,* special issue, October–December 1969. (For reply see "The Scientists Riposte," *Impact,* April–June 1970.)

OGBURN, W. F. "Technology as Environment." *Sociology and Social Research* 41 (1956), 3–9.

O'DEA, T. F. "Technology and Social Change: East and West." *Western Humanities Review*, Spring 1959.

OTHMER, D. F., and ROELS, O. A. "Power, Fresh Water, and Food from Cold, Deep Sea Water." *Science*, 12 October 1973.

PANOFSKY, W. K. H. "High-Energy Physics Horizons." *Physics Today*, June 1973.

PARKER, G. B., and DUNN, D. A. "Information Technology: Its Social Potential." *Science*, 30 June 1972.

PASSEL, PETER, and ROSS, L. "Don't Knock the Two Trillion Dollar Economy." *New York Times Magazine*, 5 March 1972.

PECHMAN, JOSEPH. "The Rich, the Poor and the Taxes They Pay," *Public Interest*, Fall 1969.

PERL, M., et al. "Public-Interest Science—an Overview." *Physics Today*, June 1974.

PERL, M. L. "The Scientific Advisory System: Some Observations," *Science*, 24 September 1971.

PI-SUNYER, ORIOL, and DE GREGORI, T. "Cultural Resistance to Technological Change." *Technology and Culture*, Spring 1964.

PLATT, JOHN. "What We Must Do," *Science*, 28 November 1969.

POOL, I. DE S. "Social Trends." *Science and Technology*, April 1968.

PREDMORE, R. L. "What Role for the Humanist in These Troubled Times?" *Bioscience*, July 1968.

PRICE, D. K. "Purists and Politicians." *Science*, 3 January 1969.

PRIMACK, JOEL. "Public Interest Science." *Science*, 29 September 1972.

RABINOWICH, EUGENE. "Back Into the Bottle?" *Bulletin of the Atomic Scientists*, April 1973.

———. "Living Dangerously in the Age of Science," *Bulletin of the Atomic Scientists*, January 1972.

———. "The Mounting Tide of Unreason." *Bulletin of the Atomic Scientists*, May 1971.

RATCHFORD, J. T. "How Scientists Advise the Congress," *Physics Today*, June 1974.

REAGAN, M. D. "Basic and Applied Research: A Meaningful Distinction." *Science*, 17 March 1967.

REED, T. B., and LERNER, R. M. "Methanol: A Versatile Fuel for Immediate Use." *Science*, December 28 1973.

REISER, L. M. The Role of Science in the Orwellian Decade." *Science*, 26 April 1974.

RICKOVER, H. G. "Can Technology be Humanized—in Time?" *National Parks*, July 1969.

RIDKER, R. G. "Population and Pollution in the United States." *Science*, 9 June 1972.

RIESMAN, DAVID. "Notes on Meritocracy." *Daedalus*, Summer 1967.

———. "Some Observations on the Limits of Totalitarian Controls." *Antioch Review* 12 (1952), 155–168.

RIVERS, CARYL. "General Turtle." *Saturday Review of Education*, May 1973.

ROBERTS, M. J. "On Reforming Economic Growth." *Daedalus*, Fall 1973.

RORVIK, DAVID. "Behavior Control: Big Brother Comes." *Intellectual Digest*, January 1974.

ROSEN, GEORGE. "Forms of Irrationality in the Eighteenth Century." In H. E. Pagliaro, *Irrationalism in the Eighteenth Century*. Cleveland, Ohio: University Press of Western Reserve, 1972.

ROSENBLOOM, R. S. "Some 19th-Century Analyses of Mechanization." *Technology and Culture*, Fall 1964.

ROSZAK, THEODORE. "The Counter Culture." *The Nation*, March 25–April 7, 1968.

ROY, RUSTUM. "University-Industry Interaction Patterns." *Science*, 1 December 1972.

RUDOFF, A., and LUCKEN, D. "The Engineer and his Work: A Sociological Perspective." *Science*, 11 June 1971.

ST. GEORGE, ANDREW. "How Does It Feel to be Bugged, Watched,

Followed, Hounded and Pestered by the C.I.A.?" *Esquire*, June 1975.

SCARF, MAGGIE. "The Anatomy of Fear." *New York Times Magazine*, 16 June 1974.

SCHOLZ, C. H., et al. "Earthquake Prediction: A Physical Basis." *Science*, 31 August 1973.

SCHLEIER, CURT. "A Campaign Against Fear." *Air Travel*, May 1970.

SCHNEIDER, MARY-JANE. "Living With Pests." *The Sciences*, December 1973.

SCHUMACHER, E. F. "Economics Should Begin With People, Not Goods." *The Futurist*, December 1974.

SCHWARTZ, CHARLES, et al. "Science and Social Controls." *Bulletin of the Atomic Scientists*, May 1969.

"Science Education: The New Humanity?" *Science News*, March 24, 1973.

SEABORG, G. T. "Science, Technology, and Development: A New World Outlook." *Science*, 6 July 1973.

SEIDMAN, AARON. "Barriers to Technical Innovation." *Bulletin of the Atomic Scientists*, March 1971.

SEVERIN, K. E. "Native Traditions—Stumbling Blocks to Progress." *Science Digest*, August 1973.

SHAPLEY, DEBORAH. "Radioactive Cargoes: Record Good but the Problems Will Multiply." *Science*, 25 June 1971.

———. "Office of Technology Assessment: Congress Smiles, Scientists Wince." *Science*, 3 March 1972.

———. "Advising the Congress: OTA Council Faces Shakedown Problems." *Science*, 9 August 1974.

SHARP, LAURISTON. "Technological Innovation and Culture Change: An Australian Case," in *Cultural and Social Anthropology*. Edited by P. B. Hammond. New York: Macmillan Company, 1964.

SHELDON, E. B., and PARKE, R. "Social Indicators," *Science*, 16 May 1975.

SHRIVER, D. W., JR. "Invisible Doorway: Hope as a Technological Virtue." *Zygon*, March 1973.

SHRYOCK, R. H. "American Indifference to Basic Science During the Nineteenth Century." *Archives Internationaux d'Histoire des Sciences* 28 (1948–49), 50–65.

SHUBIK, MARTIN. "Information, Rationality and Free Choice in a Future Democratic Society." *Daedalus*, Summer 1967.

SOLONKHIN, R. I., and BELYAEV, S. T. "Restructuring Higher Education," *New Scientist*, 3 May 1973.

SOLOW, R. "The Economist's Approach to Pollution and Its Control." *Science*, August 1971.

STARR, CHAUNCEY. "Social Benefit vs. Technological Risk." *Science*, 19 September 1969.

STUNKEL, K. R. "The Technological Solution." *Bulletin of the Atomic Scientists*, September 1973.

SULLIVAN, J. B. "Working with Citizens' Groups." *Physics Today*, June 1974.

TAYLOR, T. B., and COLLIGAN, D. "Nuclear Terrorism: A Threat of the Future." *Science Digest*, August 1974.

TELLER, EDWARD. "The Era of Big Science." *Bulletin of the Atomic Scientists*, April 1971.

THOMPSON, V. A. "How Scientific Management Thwarts Innovation." *Trans-Action*, June 1968.

THOMSEN, D. E. "The Beauty of Mathematics." *Science News*, March 3, 1973.

TOULMIN, STEPHEN. "On Teilhard de Chardin." *Commentary*, March 1965.

TOYNBEE, A. J. "The Desert Hermits." *Horizon*, Spring 1970.

———. "Not the Age of Atoms but of Welfare for All," *New York Times Magazine*, 21 October 1951.

TRIBUS, MYRON. "Technology and Society—the Real Issues." *Bulletin of the Atomic Scientists*, December 1971.

TROTTER, R. J. "Psychosurgery." *Science News*, May 12, 1973.

VON HIPPEL, F., and PRIMACK, J. "Scientists, Politics, and SST." *Bulletin of the Atomic Scientists*, April 1972.

VONNEGUT, KURT, JR. "Why They Read Hesse." *Horizon*, Spring 1970.

WADE, NICHOLAS. "Agriculture: Social Sciences Oppressed and Poverty Stricken." *Science*, 18 May 1973.

———. "Psychical Research: The Incredible in Search of Credibility." *Science*, 13 July 1973.

———. "Robert L. Heilbroner: Portrait of a World Without Science." *Science*, 16 August 1974.

———. "Theodore Roszak: Visionary Critic of Science." *Science*, 1 December 1972.

———. "World Food Situation: Pessimism Comes Back Into Vogue." *Science*, 17 August 1973.

WADDINGTON, C. H., et al. "Five Scientists View the Impacts of Technology." *Impact*, April/June 1970.

WALLACE, A. F. C. "On Being Just Complicated Enough." *Proceedings of the National Academy of Science*, January 1961.

WALLER, ROBERT. "Out of the Garden of Eden." *New Scientist*, 2 September 1971.

WALSH, JOHN. "National Science Foundation: The House That McElroy Built." *Science*, 4 February 1972.

———. "National Science Foundation: Managing Applied Research," *Science*, 11 February 1972.

WARREN, J. "Peace Pills for Presidents?" *Psychology Today*, October 1973.

WATT, K. E. F. "Man's Efficient Rush Toward Deadly Dullness." *Natural History*, February 1972.

WAYS, MAX. "It Isn't A Sick Society." *Fortune*, December 1971.

———. "Why Time Gets Scarcer." *Fortune*, January 1970.

WEAVER, W. W. "Basic Research and the Common Good." *Saturday Review*, 9 August 1969.

WEIL, ERIC. "Science in Modern Culture or the Meaning of Meaninglessness." *Daedalus*, Winter 1965.

WEINBERG, A. M. "Can Technology Stabilize World Order?" *Public Administration Review*, December 1967.

———. "Criteria for Scientific Choice." *Minerva* 1 (1963), 159–171.

———. "Criteria for Scientific Choice," "Criteria for Scientific Choice II: The Two Cultures," and "Scientific Choice and Biomedical Science." In *Criteria for Scientific Development:*

Public Policy and National Goals. Edited by E. Shils, Cambridge, Mass.: MIT Press, 1968.

WEINBERG, J. H. "Breeder Reactors. A Faustian Dilemma: Unlimited Power or Unparalled Risks?" *Science News,* 6 July 1974.

WEINER, CHARLES. "Physics Today and the Spirit of the Forties." *Physics Today,* May 1973.

WEIZENBAUM, JOSEPH. "On the Impact of the Computer on Society." *Science,* 12 May 1972.

WESTMAN, W. E., and GIFFORD, R. M. "Environmental Impact: Controlling the Overall Level," *Science,* 31 August 1973.

WHITE, HAYDEN. "The Irrational and the Problem of Historical Knowledge in the Enlightenment." in Pagliaro, *Irrationalism in the 18th Century.* Cleveland, Ohio: University Press of Case Western Reserve, 1972.

WHITE, LYNN, JR. "The Historical Roots of Our Ecological Crisis." *Science,* 10 March 1967.

WHYTE, L. L. "Science and Our Understanding of Ourselves." *Bulletin of the Atomic Scientists,* March 1971.

WIENER, NORBERT. "Some Moral and Technical Consequences of Automation," *Science,* 6 May 1960.

WILLIAMS, GURNEY, III. "The Use and Abuse of Fear Psychology." *Science Digest,* February 1975.

WILSON, J. Q. "Why We Are Having a Wave of Violence." *New York Times Magazine,* 14 May 1968.

WINTHROP, HENRY. "Some Roadblocks on the Way to a Cybernated World." *American Journal of Economics and Sociology,* October 1966.

WOLMAN, ABEL. "The Metabolism of Cities." *Scientific American,* September 1965.

WORSLEY, P. M. "Cargo Cults," *Scientific American,* May 1959.

YOUNG, GORDON. "Whatever Happened to TVA?" *National Geographic,* June 1973.

ZECKHAUSER, RICHARD. "The Risks of Growth." *Daedalus,* Fall, 1973.

Fiction:

BUTLER, SAMUEL. *Erewhon.* Various editions.

CAPEK, KARL. *R.U.R.* New York: Oxford University Press, 1961.

CHARNAS, SUZY. *Walk to the End of the World.* New York: Ballantine Books, 1974.

CLARKE, A. C. *Childhood's End.* New York: Ballantine Books, 1963.

DICKENS, CHARLES. *Bleak House.* Various editions.

DICKENS, CHARLES. *Hard Times.* Various editions.

GRAVES, ROBERT. *Watch the North Wind Rise.* New York: Creative Age Press, 1949.

HAWTHORNE, NATHANIEL. "The Celestial Railroad." In *The Complete Short Stories of Nathaniel Hawthorne,* Garden City, N.Y.: Doubleday, 1959.

HORNE, R. H. *The Great Peacemaker. A Sub-Marine Dialogue.* London: Printed for private distribution, 1871.

HUXLEY, ALDOUS. *Brave New World.* Various editions.

JOHANNESSON, OLAF. *The Great Computer.* London: Gollancz, 1968.

JONES, D. F. *Denver Is Missing,* New York: Berkeley Medallion, 1974.

LeGUIN, URSULA. *The Left Hand of Darkness,* Ace Books, 1969.

MELVILLE, HERMAN. "The Tartarus of Maids." *The Complete Stories of Herman Melville,* New York: Random House, 1945.

MERLE, ROBERT. *Malevil.* New York: Simon & Schuster, 1973.

ORWELL, GEORGE. *1984.* Harcourt, Brace, and World, 1949, 1962.

PYNCHON, THOMAS. *Gravity's Rainbow.* New York: Viking Press, 1973.

SKINNER, B. F. *Walden Two,* New York: Macmillan Company, 1948.

SPENCER, SCOTT. *Last Night at the Brain Thieves Ball.* Boston: Houghton Mifflin Company, 1973.

STAPLEDON, OLAF. *Last and First Men.* London: Methuen & Co., 1931; New York: Dover, 1968.

SZILARD, LEO. "The Mark Gable Foundation." In *The Voice of the Dolphin*. New York: Simon and Schuster, 1961.

VONNEGUT, KURT. *Player Piano*, New York: Holt, Rinehart & Winston, 1952.

WORDSWORTH, WILLIAM. "On the Projected Kendal and Windermere Railway." 1844. (In various collections.)

Reports, Booklets, Cassettes, Papers:

ARON, RAYMOND. *The Epoch of Universal Technology*. London: Encounter Pamphlets, 1964.

BORETSKY, MICHAEL. *U.S. Technology: Trends and Policy Issues*. Washington, D.C.: George Washington University, October 1973. (Available from National Technical Information Service.)

BROWN, B. S., *et al. Psychosurgery. Perspective on a Current Problem.* National Institute of Mental Health, 1973 (DHEW # (HSM) 73–9119).

COATES, VARY T. *Technology and Public Policy; The Process of Technology Assessment in the Federal Government*. 2 vols. plus summary report. Washington, D.C.: George Washington University, July 1972.

Committee on Inter-governmental Science Relations. *Public Technology, a Tool for Solving National Problems*. Washington, D.C.: U.S. Government Printing Office.

Environmental Protection Agency. *The Quality of Life Concept. A Potential Tool for Decision-Makers*. 1973.

Joint Economic Committee. *Federal Transportation Policy: The SST Again*. Report of the Subcommittee on Priorities and Economy in Government, 1973. (Available from U.S. Government Printing Office.)

LAING, R. D. *Politics as Experience*. San Fernando, California: Superscope Educational Products (cassette).

MALENBAUM, WILFORD. *Materials Requirements in the United States and Abroad in the Year 2000*. Washington, D.C.: National Commission on Materials Policy, 1973.

MAY, ROLLO. *Violence and Spirituality.* San Fernando, California: Big Sur Recording (cassette).

National Academy of Sciences, Committee on Science and Astronautics. *Technology: Processes of Assessment and Choice,* Washington, D.C.: U.S. Government Printing Office, July 1969.

National Science Foundation. *The Effects of International Technology Transfer on U.S. Economy.* Papers and Proceedings of a Colloquium held on November 17, 1973 (NSF 74–21). July 1974.

———. *Research and Development and Economic Growth (Productivity).* Papers and Proceedings of a Colloquium (NSF 72–303). 1972.

Office of Technology Assessment and Forecast. *Technology Assessment and Forecast.* Washington, D.C.: U.S. Department of Commerce, May 1973 and December 1973.

RASKIN, EUGENE. *Life in the City.* North Hollywood, California: Center for Cassette Studies.

ROTHSCHILD, JOAN A. "A Feminist Perspective on Technology and the Future of Human Society," Paper presented at the Second General Assembly of the World Future Society, Washington, D.C., June 2–5, 1975.

Science and Technology Policy Office. *Chemicals and Health.* Report of the Panel on Chemicals and Health of the President's Science Advisory Committee. Washington, D.C.: National Science Foundation. September 1973.

TOULMIN, STEPHEN. . . .*and Shall We Have Science For Ever and Ever?* American Association for the Advancement of Science, Washington, D.C.: (cassette).

INDEX